The Evolution of Complex Regional Pain Syndrome

Michael Stanton-Hicks

The Evolution of Complex Regional Pain Syndrome

From Schloss Rettershof to a New Clinical Language

Michael Stanton-Hicks
Department of Pain Management
Cleveland Clinic
Cleveland, OH, USA

ISBN 978-3-031-54902-1   ISBN 978-3-031-54900-7   (eBook)
https://doi.org/10.1007/978-3-031-54900-7

This Springer imprint is published by the registered company Springer Nature Switzerland AG
The registered company address is: Gewerbestrasse 11, 6330 Cham, Switzerland

If disposing of this product, please recycle the paper.

Foreword

My friend and colleague Michael Stanton-Hicks has become the recorder of record of the CRPS story: every detail of the long and tortuous tale of CRPS as it evolved from reflex sympathetic dystrophy (RSD) has been recorded and put before the reading public. This has been a task that has lasted over 50 years, and the final chapter has not yet been written. One can only hope that it is completed while Michael is still with us so that we will have his fully detailed record.

The story begins with the meeting that led to the founding of IASP, held in Issaquah, Washington, and planned and implemented by John J. Bonica, MD. A small, self-selected group determined that the existing terminology based upon reflex sympathetic dystrophy (RSD) needed to be updated and the clinical picture as well as the basic science needed to be clarified. Michael was the organizer of the group that met at Schloss Rettershof in 1988, after 15 years of discussion and correspondence. This meeting, by invitation only, was clearly the most important in the series that has revised our understanding of the clinical condition and the mechanisms underlying it. The plan for a Special Interest Group (SIG) of IASP was launched, and all subsequent meetings were held under this rubric. Five years later a second meeting was held in conjunction with the American Pain Society Meeting in Orlando, FL. This meeting created the concept that CRPS was a variant of neuropathic pain that involved nociceptive and inflammatory processes.

Five years later a meeting was held in Orlando, FL, in 1993 to work on the development of diagnostic criteria, and 4 years after that a meeting in Malibu, CA, prepared research plans and a new diagnostic algorithm. Two meetings quickly followed, the first in Cardiff that was a satellite of the IASP Congress in Glasgow that focused upon animal experimental data and the second in Budapest that followed a World Institute of Pain (WIP) meeting in 2003 that was considered a meeting to develop a consensus on all things CRPS. The document developed by this group was forwarded to the IASP Committee on Taxonomy in 2005.

Chapter 7 discusses case management including pediatric CRPS, which has several features that do not occur in adults, and Chap. 8 introduces the concept of biomarkers as a means of establishing the diagnosis. The CRPS group of researchers and clinicians succeeded in getting this diagnosis of chronic pain introduced into the ICD Classification of Diseases: a major triumph.

This entire process, from 1973 until the present day, was mainly the work of Michael Stanton-Hicks, who made it the principle focus of both his research and clinical activities. We owe him a great debt for bringing the work of this group to the scientific and medical worlds and providing us with a narrative of every step in the process.

John D. Loeser
Neurological Surgery
University of Washington
Seattle, WA

Preface

The aim of this book is to tell the story of how pain practice, research, and health-care, in the case of Reflex Sympathetic Dystrophy, can benefit from a revision of its nomenclature. It is hoped that this volume will be of interest for pain practitioners, researchers of pain or related mechanisms, and other allied scientists whose activities touch on the science of pain and its management. The text describes how an inadequate medical term underwent structural changes that more adequately reflect the nature of its pathophysiology. This was achieved by the efforts of an extraordinary group of interested clinicians and medical scientists. Seeds for this work were already laid in Seattle where plans for an initial meeting that would ultimately be held at a small castle, Schloss Rettershof, Kelkheim near Mainz, Germany, in conjunction with the Annual Meeting of the European Society of Regional Anaesthesia were made.

Early discussions regarding its content with colleagues on the faculty of the University of Washington Department of Anesthesiology coincided with plans by Dr. John Bonica for the Inaugural International Pain Meeting at Issaquah WA, 1974, at which the IASP was founded. I should also acknowledge that Louisa E. Jones, as John Bonica's secretary, was intimately involved in the organization of that Meeting and who ultimately became the Executive Secretary of the IASP, a position to which she was devoted until her retirement in 2005. I am indebted to Louisa and the IASP office for the early advice and subsequent help I received over the years during which and for all the subsequent meetings of the IASP Special Interest Group (SIG) Pain and the Sympathetic Nervous System under which auspices the development of the new taxonomy constitute the theme of this text.

Because of the time scale and large number of faculty involved, in addition to its expected readership, I decided to make this a photographic record that would put a face to the contributors, all of whom gave up so much of their time to achieve such a successful outcome. This has been a momentous challenge although as an avid photographer, I have a large inventory of Leica portraits of colleagues to which I have added many photos that I requested from participants and friends of faculty many of whom have since passed away. For some of these images, I would like to thank Louisa Jones, John Loeser, Laurie Mather, Steve Abram, Stephen Bruehl, Stephen Butler, Frank Birklein, Angela Mailis, Gary Bennett, Wilfrid Jänig, Hans Nolte, Hermann Kreuscher, Louise Anne Oaklander, Philip Finch, Gunner Olsson, Norm Harden, and Christopher Glynn.

Chapter 1 describes the groundwork that was the inspiration for all subsequent meetings and in many respects was the most consequential workshop of all as it incorporated all the doubts, points of view, and enthusiasm from a bunch of highly experienced scientists and clinicians, all of whom acknowledged the serious shortcomings that were associated with its current term Reflex Sympathetic Dystrophy (RSD). Undoubtedly, my role as organizer was lightened as the success of this inaugural meeting really rested on the shoulders of three individuals Wilfred Jänig, Manfred Zimmerman, and Hans Nolte—assisted by Albert van Steenberg and Hermann Kreuscher who managed to keep the show on the rails so to speak such that by the end of the workshop a document which embodied the skeleton of a new terminology was already available for submission to the IASP together with the request that the SIG Pain and the Sympathetic Nervous System as a new Special Interest Group in the IASP should be incorporated. While no proceedings of this meeting were recorded, I am very grateful for Prithvi Raj who as the editor of the Current Management of Pain series invited me to include the transcripts I had collected from the meeting for publication as a book entitled, *Reflex Sympathetic Dystrophy*, Kluwer Academic Publishers, 1990.

The next chapter in this story took place 5 years later in Orlando in conjunction with the American Pain Society (APS) Annual Meeting. Both the organizers of the APS and the IASP office were very helpful in selecting a venue and providing logistic support for the workshop. The theme of this meeting was quite different from that at Schloss Rettershof in that a more enlightened perspective of the manner in which the SNS system interacts in the pathology of RSD and causalgia prevailed; it was seen rather as a component in the neuropathic, nociceptive, and inflammatory processes that constitute the pathophysiology of these two clinical entities, not necessarily the driver. For this I am particularly indebted to Wilfrid Jänig whose indelible experience with the autonomic nervous system prevailed like a *leitmotiv* throughout all the workshops.

Many of the faculty for this workshop were chosen because of their individual research into scientific and clinical aspects of RSD pathophysiology. Chapter 3 describes the 2-day meeting that was held in Malibu 1997. Fine tuning of the diagnostic criteria incorporated much of the dialog. The most important aspect, however, concerned validation of the data, and a multicenter international study overseen by Norman Harden subsequently determined the statistical credibility of the diagnostic criteria.

As an interjection to the flow of CRPS diagnostic criteria development, the very welcome and pertinent IASP international, multidisciplinary, Research Symposium was held in Cardiff in conjunction with the IASP Congress in Glasgow, 2000. We are indebted to the early efforts of Bradley Galer who conceived the idea of showcasing current research of CRPS pathophysiology.

Chapter 5 describes the meeting in Budapest which was the culmination of 15 years work that saw the production of a draft of the proposed clinical diagnostic criteria for CRPS that was submitted to the IASP for consideration by the Committee of the Classification of Chronic Pain to replace the existing definition. The conundrum of still having only a sketchy knowledge of CRPS pathophysiology was an impediment to the development of adequate diagnostic criteria, and the specific measures that were taken to ensure its success are well described within the text. As can be appreciated, the statistical methodology that was used to validate and improve both research and clinical utility of the Criteria was realized. This was also the largest attended workshop, and special acknowledgment and thanks go to Gabor Racz and Dr. Edit Racz, Blaguss Congress Bureau, Sandra Vamos, on-site organizers of the World Institute of Pain Congress (WIP) that was convened at the same time in Budapest. I am indebted to their unstinted help in the logistics of running our SIG Workshop and moving faculty between the airport and the two venues. Special thanks are also extended to many pharmaceutical companies, Medtronic Corporation, B. Braun Medical, Astra Pharmaceutical, Pharmacia-Deltec, Baxter Health Care Corp., Burron Medical, and Epimed who provided unrestricted grants without which the meeting could never have taken place.

The remaining three chapters are a postscript to the introduction of the new terminology Complex Regional Pain Syndrome. Chapter 6 describes the influence that the term CRPS has had on clinical practice, basic science research, and the practice of pain as viewed by other medical specialties and especially organized medicine such as the Veterans Administration System, Centers for Medicare and Medicaid Services (CMS), American Medical Association (AMA) in the United States, and the World Health Organization (WHO)—International Classification of Medical Diseases (ICD-11).

A collection of adult and pediatric CRPS cases in Chap. 7 emphasizes the sometimes extraordinary treatment and rehabilitation measures that may be required to achieve the best possible clinical outcome. Particularly stressed is the fact that many cases, in particular children, will respond to early simple measures that will frequently achieve a complete remission within a few weeks, while other more complicated cases can require a lifetime investment in the management of CRPS symptoms.

The final chapter is a window on the application of biomarkers to the diagnosis of medical diseases including CRPS. While a biomarker can be a specific link to an aspect of the pathophysiology of a disease, in the case of CRPS, the diagnostic credibility may rest on not one but several such characteristics that have a temporal signature or may be predictive of a response to treatment.

I would like to acknowledge several individuals, without whose encouragement and support this work might not have materialized. Andre Machado, MD, neurosurgeon, a colleague with whom I have worked for many years, was particularly insistent that I document the story of CRPS. I have been equally encouraged by my good friend and seer, Joshua Prager, MD, in the realm of sympathetic disorders, a domain we have shared for many decades. I would be very remiss if I did not mention Robert Boas MB; BS, friend, colleague, and avant garde from the Seattle days who is equally part of this tale. I am particularly grateful for the support, advice, and incentive I received from John Loeser, MD, Seattle denizen, friend and colleague who accepted the onerous task of reviewing the manuscript. Others who have been equally consequential to the theme of this work are Samuel Hassenbusch, MD, deceased, wonderful friend and colleague with whom I worked for many years, and

Mark Hendrickson, MD, colleague and friend whose interest and skill in peripheral nerve stimulation surgery has helped to relieve the symptoms of many a CRPS patient.

Finally, this work is dedicated to John Bonica, MD, who awakened the world to the need to relieve pain and who not only started me on this journey but has been my beacon ever since.

Cleveland, OH, USA   Michael Stanton-Hicks

Prologue

Long before any written accounts of what has come to be described as Complex Regional Pain Syndrome (CRPS), this book entails a record that began in the department of Anesthesiology headed by Dr. John Bonica who during the Pacific arena of World War II came to manage thousands of wounded servicemen many of whom suffered from this complicated clinical entity often long after their original wounds had healed.

For millennia, injury has shaped life. By eliminating the unfit, memory of such events shapes the tissues and cells, usually in the form of life-saving reactions that can be triggered at a moment's notice. CRPS usually follows some form of trauma which even to the medically trained often appears to be an inappropriate response by the central nervous system, the peripheral nervous system, the autonomic nervous system together with an inflammatory reaction that frequently seems to have both autoimmune and genetic characteristics. While an incident associated with the onset of CRPS can generally be identified, it is frequently minor and even some cases can arise de novo. No doubt this entity has been described by witchdoctors, apothecaries, and other medical and non-medical men for millennia. While this clinical disorder has been given many names, for the sake of uniformity the term Reflex Sympathetic Dystrophy (RSD) will be used throughout the early part of this account until the first Consensus Meeting of medical practitioners, basic scientists, and behavioralists at Schloss Rettershof in 1988 was held to review the available data and see if a common terminology could be developed that would be used as the basis for a new taxonomy. It was the hope of all who attended that a new term would be proposed to the International Association for the Study of Pain (IASP). After its adoption by IASP, the current acronym Complex Regional Pain Syndrome (CRPS) will be used throughout the remainder of the text. Although there have been a lot of individual reports that describe the symptoms of extreme burning pain that can occur as a consequence of nerve injury during the past half century, the superb account by Silas Weir Mitchell and colleagues in their book *Gunshot Wounds and Other Injuries of Nerves* that was published during the American Civil War must remain the first complete and elegant description of Causalgia or what is now referred to as CRPS 2, Mitchell et al. (1864) (Table 1). It is interesting to reflect on the observations of other physicians like Denmark who some 50 years before Mitchell during the Peninsular War (1804–1807) described amputation of the arm of a soldier due to burning pain in the median nerve distribution: Paget (1814–1899) had also noted the "distressing and hardly imaginable pain and disability in the fingers of some patients with injured nerves." Buried in his extensive collection of papers is a detailed description of the inflammatory response that occurred in the arm of King Charles of France by Ambroise Paré (1510–1590) who as surgeon to the king had on one occasion encountered nerve structures during bloodletting for the treatment of smallpox. This account some 200 years earlier than Mitchell adds to what we know about the response of a nerve to what is in reality a contusion injury and one which only seems to rarely occur after resection of a nerve. Many early reports typically described the passage of a musket shot in the vicinity of a major nerve often with little collateral damage. In earlier times, the same response would very likely have occurred after passage of an arrowhead! The incidence of causalgic injuries during the Civil War have been reported to be as high as 38% (Echlin et al. 1949). Very few reports are available from World War I. Carter (1922) described 23 cases of major or minor causalgia in 1020 nerve injury cases. However, as can be seen from Table 1, the incidence of causalgia reported varied between 2% and 20%.

Interest in studying this phenomenon which was not only seen after nerve injury but would also occur after relatively mild blunt trauma or following a fracture was characterized with even greater frequency. Leriche during World War I described his surgical approach to dissection of sympathetic nerve fibers in a distal extremity to combat burning pain. The Second World War galvanized several surgeons to not only study the phenomenon but also develop surgical methods to address nerve pathology. Foremost amongst these were, Kirklein et al. (1947), Echlin et al. (1949), Ulmer and Mayfield (1946), Nathan (1947), and the Australian surgeon/neurologist, Sir Sidney Sunderland (1991). Interestingly, like Mitchell they all noted some altered relationship (dysregulation) of sympathetic function which when blocked either centrally at the ganglia with local anesthetics or by distal surgical sympathectomy (Leriche 1916) seemed to immediately relieve the burning symptoms. These views hark back to Claude Bernard, that great French physiologist, Bernard (1858), who had already demonstrated a role of the sympathetic nervous system in the generation of pain. Bonica (1973), Carron and Littwiller (1975), and Evans (1946) all promoted local anesthetic sympathetic blocks to manage such symptoms in patients. In 1946, Evans introduced the term Reflex Sympathetic Dystrophy (RSD) to the

Table 1 Incidence of causalgia

	Nerve injuries	Causalgia cases	Incidence %
Spiegel and Milowski	275	9	3.3
Ulmer and Mayfield	1477	75	5.1
Nathan	160	22	13.8
Kirklein et al.	2850	52	1.8
Freeman	2167	114	5.0
Slessor	670	28	4.2
White et al.	400	13	3.3
Sunderland and Kelly	282	34	12.0
Echlin et al.	1500	30	2.0

Data are taken from servicemen of the American Civil War who had sustained nerve injuries that were associated with the development of Causalgia. Richards RL. Causalgia: a centennial review. *Arch Neurol.* 1967:16;339–50 (reprinted with permission)

lingua franca and firmly anchored the relationship of "burning symptoms" with disordered autonomic sympathetic function.

Paul Sudeck is also extremely important to this story as he was the first to describe the excessive inflammatory response and bone lesion that was invariably associated with RSD. These observations are documented radiographically and in detailed writings, Sudeck (1942), Van der Laan and Goris (1997), Köster et al. (2012). The disease entity is also known in Europe as Morbus Sudeck.

While there is no better account of causalgia than that by Richards (1967), "Causalgia: A Centennial Review," the case for RSD which occurs with far greater frequency, particularly during peacetime, is my main excuse for relating its convoluted course and to identify the disease entity under a common term that implies no mechanism and only acknowledges its main clinical features. The path for unraveling this tale commences below.

References

Bernard C. Lecons sur la Physiologie et la Patho;ogie du Systeme Nerveux, 2vols, Paris: Bailliere. 1858.

Bonica JJ. Causalgia and other reflex sympathetic dystrophies. Postgrad Med. 1973;53:143–8.

Carron H, Littwiller R. Stellate ganglion block. Anesth Analg. 1975;54:567–70.

Carter HS. On causalgia and allied painful condition 3: due to lesions of peripheral nerves. J Neurol Psychopathol. 1922;3:1–38.

Echlin F, Owens FM, Wells WI. Observations on "Major" and "Minor" causalgia. Arch Neurol Psychiatr. 1949;62:183203.

Evans JA. Reflex sympathetic dystrophy. Surg Clin North Am. 1946;26:78–90.

Kirklein JW, Chenoweth AI, Murphy F. Causalgia: a review of its characteristics, diagnosis and treatment. Surgery. 1947;21:321–42.

Köster U, Stanton-Hicks M, Maihöfner C. What Paul Sudeck already suspected: from Sudeck's disease to complex regional pai syndrome. Schmerz. 2012;26:438–40.

Leriche R. De la causalgia comme envisege'e une ne'vrite du sympathique et son traitement par la de'nudation et l'excision des plexus nerveux periarteriels (in French). Presse Med. 1916;24:170–80.

Mitchell SW, Morehouse GR, Keen WW. Gunshot wounds and other injuries of nerves. Philadelphia: JB Lippincott; 1864.

Nathan PW. On the pathogenesis of causalgia in peripheral nerve injuries. Brain. 1947;70:145–70.

Richards RL. Causalgia: a centennial review. Arch Neurol. 1967;16:339–50.

Sudeck P. Die sogenannte akute Knochenatrophie als Entzündingsvorgang. Chirurg. 1942;14:449–58.

Sunderland C. Nerve Injuries and tyheir Repair: A Critical Appraisal. Churchill Livingstone, Edinburgh. 1991.

Ulmer JL, Mayfield FH. Causalgia; a study of 75 cases. Surg Gynecol Obstet. 1946;83:789–96.

Van der Laan L, Goris RJA. Sudeck-syndrom: hates Sudeck recht? Unfallchirurg. 1997;100:90–9.

Acknowledgments

This book is a tribute to my patients whose fortitude has taught me so much about their courage in the face of disability. I must also acknowledge my wonderful wife, Ursula, without whose understanding and patience the book would never have been completed. Finally, I am ever grateful to my father whose wise council guided me into the scientific underpinnings of the medical profession.

Contents

List of Figures

List of Tables

1

1.1   The Influence of Schloss Rettershof

My own interest in this clinical entity began during the 5 years that I worked in the Department of Anesthesia at the University of Washington and Harborview Medical Center in Seattle that were chaired by John Bonica, MD. I soon found out that this would be an exercise of immersion as already mentioned above. John Bonica had literally taken care of thousands of patients who had developed RSD after their various injuries during the recent War. Use of sympathetic blocks was ubiquitous and for a shorter or longer duration would provide considerable pain relief for most cases. My friends and colleagues, many expatriates from British Commonwealth countries, also working in the same department could not but help share the same excitement and interest in this weird condition.

fr. l. to r., Terrence "Terry" Murphy, Louis Jacobsen, John Bonica, Ed Charlton

"Will" Fordyce

Before continuing with my RSD/CRPS story which as stated above began in John Bonica's department, I will mention some "pre-history" of RSD vascular research which was undertaken in Kiel and Freiberg, Germany respectively by Wilfrid Jänig and Helmut Blumberg—see Appendix. I would like to share some of the practical aspects as they affected our duties of providing anesthesia and pain services in what at the time was in fact the first multidisciplinary pain clinic in the country. Most of my anesthesia duties were carried out at Harborview Medical Center with some rotation to the Veterans Administration Medical Centre and the University Medical Center. Attendance at the very important early morning meetings at which both basic science members of the research team and clinicians met to discuss different aspects related to management issues or "hot button" topics such as RSD was required. In this context, management with sympathetic blocks, other sympatholytic measures, medications or psychological instruments and of course *Operant Conditioning* introduced by Wilbert (Bill) Fordyce in the Department of Physical Medicine and Rehabilitation, Fordyce et al. (1973) was *de rigeur.* This latter technique was quite a new approach to the behavioral management of chronic pain and as was clear in the clinical setting it turned out to be a highly effective adjunct to help those cases in which significant psychological issues predominate.

Arthur Ward

Louisa Jones and John Bonica

Not infrequently, Arthur Ward, Chair of the Department of Neurosurgery or John Loeser from the same department and "Bill" Fordyce from the Department of Physical Medicine and Rehabilitation would attend or be invited to speak at these discussions not only related to practical administrative aspects of pain or case management issues but also new innovations in the treatment of chronic pain. Characteristically, there was a natural division in the conference room in that all the clinicians sat on one side and the researchers, and their technicians sat on the other side of the room, the twain would only mingle during some of the often-animated discussions always moderated by John Bonica—when in town—or Tom Hornbein (vice chair) or one of the other senior staff. During 1972, many debates were held relating to the material that Dr. Bonica would like to see covered in the planned inaugural International Pain Conference for basic scientists and clinicians that would eventually be held in May 1973 at the Sisters of Providence Training Centre (which had not long before changed its function as a training center for novices of the Order to that of the Providence Heights Conference Centre) in Issaquah WA (Fig. 1.1). While John Bonica was the prime instigator,

Fig. 1.1 The Providence Heights Conference Center. Site of the inaugural International Pain Conference, Birthplace of IASP

organizer, promotor, and executor of the meeting I was aware that Louisa Jones who as Director of Research and Publications in the Department of Anesthesiology at the University of Washington also helped in its organization. At the behest of John Bonica, she subsequently became Executive Director of a nascent IASP society born in Issaquah, 1973 which she coddled during its growth and over the many successive years of its mission. The remarkable aspect of the International Symposium is that 339 delegates, over 70% of which were from other parts of the United States or overseas, made up the largest assembly of pain basic scientists and clinicians, many of whom did not even know one another beforehand. John Bonica in his inimitable fashion collected his speakers (102) from his reading of the literature and asking those whom he knew to suggest others who would address topics that he wanted debated at the meeting. Funding for the meeting was ampul, much forthcoming from the National Institutes of Health, the Dean of the University of Washington, Robert van Citter's and many pharmaceutical companies. Because of its isolation, all the participants were essentially domiciled for 5 days in quite primitive sparse rooms originally intended as a dormitory for Sisters of Providence novices: everyone was so excited about the event that there were no complaints!

"Steve" Butler

When we had time on our hands which was not often, or off campus over a beer and local salmon gustation, the topic RSD and its management occupied the attention of our tightly knit group, Robert (Bob) Boas, Terry Murphy, Steve Butler, Ed Charlton and me. Most of our research was carried out on human volunteers and was related to the pharmacokinetics of local anesthetics like bupivacaine and etidocaine, both long-acting local anesthetics and potentially very useful in many regional anesthetic procedures that would call for symptom control while managing the functional restoration of patients suffering from the consequences of RSD (Fig. 1.2). In fact, one of the agents, bupivacaine has become the standard component in the prolonged epidural infusion of analgetic solutions for pain management of CRPS patients undergoing their rehabilitation. Our colleagues Geoffrey Tucker and Laurie

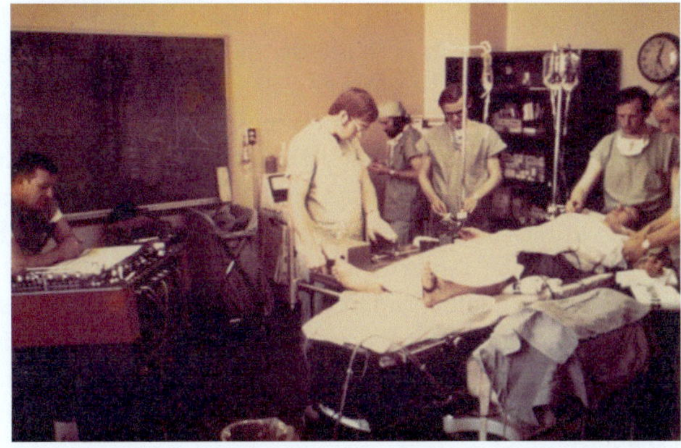

Fig. 1.2 Human research: from L to R, Charles "Chuck" Percy, "Laurie" Gary LeDray Laurie Mather, Geoff Tucker, Terry Murphy, Michael Stanton-Hicks

Mather, from the University of Washington and the Virginia Mason Hospital, both highly experienced pharmacokineticists were our constant allies and friends during this quite detailed research.

Laurie Mather and Goeff Tucker

John Liebeskind and Ron Melzack who had visited the department on several occasions were each empathetic and very supportive of our group's intentions to further explore the syndrome reflex sympathetic dystrophy. So, when my time in the department was coming to an end and I was moving to a new position at the University of Massachusetts Medical Center, Worcester several of us including Steve Butler, Terry Murphy, Robert Boas and Ed Charlton all agreed to stay in touch with a view to holding a meeting/workshop on RSD at one of the IASP or Chapters of the American Pain Society meetings in the future. Bob Boas and Laurie Mather later joined me in Worcester to help set up research and a Pain Clinic in the image of the University of Washington model. Bob Boas, of course who had a special interest in RSD would become an important part of the first meeting at Schloss Rettershof. As is typical for such endeavors, this would have a long gestation.

John Liebeskind and Ron Melzack

Ulf Lindblom and Heinrich Fruhstorfer

The first such opportunity was some 15 years later in conjunction with the 1988 European Society of Regional Anaesthesia meeting, at the Johannes Gutenberg University in Mainz, Germany where I held a position on their faculty. The venue for the Workshop was a small castle, Schloss Rettershof, in nearby Kelkheim. All the old Seattle faculty together with David Haddox and Steve Abram from the Medical College of Wisconsin, Manfred Zimmermann, Abteilung für Physiologie, Heidelberg Universität and Wilfrid Jänig Physiologisches Institut, Christian-Albrechts Universität, Kiel, Germany and many others made up the list of attendees—shown in Table 1.1. The faculty composition from nine countries represented Clinicians and Basic Scientists whose medical disciplines or routine research touched on sympathetic nervous system dysregulation as being one of the principal components underlying the symptom complex RSD (Fig. 1.3).

Table 1.1 Participants and their affiliations at Schloss Rettershof are shown in the table

Stephen E Abram, MD	H Kruescher, Dr. med
Depart of Anesthesiology	Institut für Anaesthesiologie, Stadt Kliniken
Medical College of Wisconsin	Osnabruck, FRG
Helmut Blumberg, Dr. med	H. Nolte, Dr. med
Abteilung Für Klinische Neurologie	Abteilung für Anaesthesiologie
Klinikum der Albert-Ludwigs Uni. Freiburg	Kresikrankenhaus, Minden, FRG
Helmut Blumberg, Dr. med	
Abteilung für Neurologie	
Albert-Ludwigs Universität, Freiburg	
Robert A. Boas MB;BS	Jennifer Kelly PhD
Department of Clinical Pharmacology	University Center for Pain Medicine at Hermann
Uni. of Auckland, New Zealand	University of Texas

(continued)

Table 1.1   (continued)

Stephen H Butler MD Department of Anesthesiology University of Washington, Seattle	Royce Lewis MD Department of Plastic Surgery University of Texas
Jeffrey Cannella MD Center for pain medicine at Hermann University of Texas	Terrence Murphy MB;BS Department of Anesthesiology and Clinical Pain Service University of Washington, Seattle
Edmond J. Charlton MB; BS Department of Anaesthetics University of Newcastle, England	O Nickel Dr. med Institut für Strahlenkunde, Abt. für Nuklearmedizin Johannes Gutenberg Universität. Mainz FRG
J. David Haddox DDS Department of Anesthesiology and Psychiatry Medical College of Wisconsin, Milwaukee	Gabor B. Racz MD Department of Anesthesiology University of Texas
H Fruhstorfer, PhD Institut für Normale und Pathologische Physiologie Phillips-Universität, Marburg, FRG	P. Prithvi Raj MD University Center for Pain Medicine at Hermann University of Texas
U Gebershagen, Dr. med Schmerz Zentrum, Mainz, FRG	Michael Stanton-Hicks MB; BS, Dr. med Klinik für Anaesthesiologie
K. Hahn MD Inst. für Strahlenkunde, Abt. für Nuklearmedizin Klinik. der Johannes-Gutenburg Uni. Mainz, FRG	Johannes Gutenberg Universität, Mainz, FRG Ronald Tasker, MD Division of Neurosurgery
JG Hannington-Kiff MSc, MB; BS Pain Relief Center, Frimley Park Hospital, England	Toronto General Hospital, Canada Eric Torebjörk MD, PhD
James E. Heavner PhD Department of Anesthesiology University of Texas	Department of Clinical Neurophysiology University Hospital, Uppsala, Sweden Albert van Steenberg, Dr. med
ME Hornyak Dr. med Abt. für Klinische Neurologie and Neurophysiologie Klin.der Albert-Ludwigs Universität, Freiburg, FRG	Department of Anaesthesia Klinikum St Anne, Bruxelles, Belgian Peter R. Wilson MB, BS, PhD
Wilfrid Jänig, Dr. med Physiologisches Inst. Christian-Albrechts Uni. Kiel, FRG	Department of Anesthesiology Mayo Clinic, Rochester, USA
Ilmar Jurna, Dr. med Inst. für Pharmakologie und Toxokologie Universität des Saarlandes, Homburg, FRG	M. Zimmerman, Dr. med Physiologisches Institut, Universität Heidelberg, Heidelberg, FRG

Fig. 1.3 Schloss Rettershof, Kelkheim, Germany, conference room and front cover of the program where the inaugural consensus workshop was convened in October 15–17, 1088

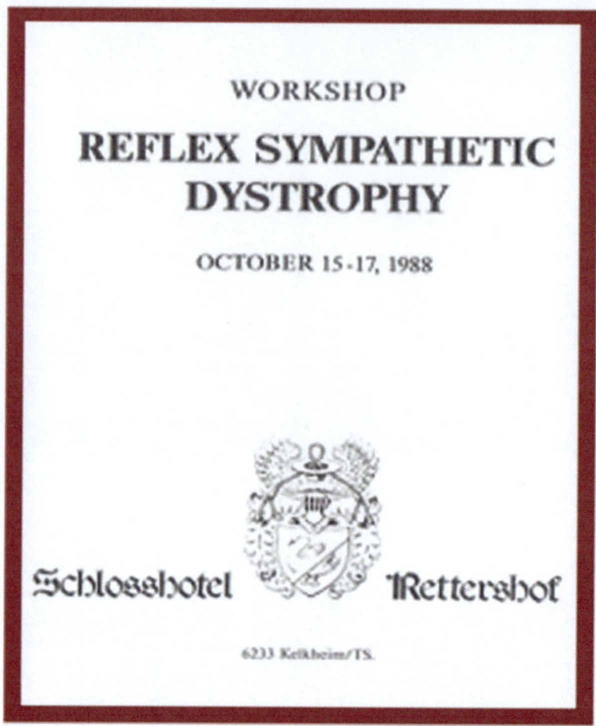

Fig. 1.3 (continued)

1.2   The Charge

Schloss Rettershof provided an ideal isolated environment to hold our Workshop. A dedicated staff who remained almost invisible throughout the Workshop catered to all the audio-visual, logistic aspects and meals during the 2½ days occupied by the Workshop. By its nature It was conducive to an extremely intense, immersive, and productive experience to the point that practically all of our expectations remarkably, were achieved. The meeting would continue discussions that had begun in Seattle 15 years earlier together with an expanded international group of interested parties to clearly: (1) define the syndrome of RSD, (2) develop a proposal to the Task Force on Taxonomy of the IASP and (3) to reclassify the definition of the clinical entities, Causalgia and RSD. It was our hope and expectation that a new classification and Terminology would appear in the next edition of the Classification of Chronic Pain: Descriptions of Chronic Pain Syndromes and Definitions of Pain Terms, IASP press. In addition to the foregoing, our belief that by using such an approach we could provide a uniform platform for clinicians to use and ultimately set the course for basic and clinical research into the patho-physiology and epidemiology of these disorders that should ultimately provide an improvement in patient care.

Three key faculty, Wilfrid Jänig, Dr. med, PhD, the Doyen of autonomic research—Physiologisches Institut, Christian Albrechts-Universität, Kiel, FRG, Manfred W. Zimmerman, PhD, Physiologisches Institut, Universität, Heidelberg. FRG and Hans Nolte, Dr. med, Chefarzt, Abteilung für Anaesthesiologie, Kreis Krankenhaus, Minden, FRG, together with Albert van Steenberge MD, Department of Anaesthesia, Klinik St. Anne, Bruxelles, Belgium and H Kreuscher, Dr. med, Institute für Anaesthesiologie, Stadt Kliniken, Osnabruck, FRG. were discussants.

Wilfrid Jänig

1.3 First Day

1.3.1 Session 1. Opening and Agenda

An introduction of the meeting that included a palette of the convoluted history of RSD, the shortcomings for patient care and various attempts to study its current chaotic diagnosis, pathophysiology and techniques used for its topical management comprised the opening.

This was structured to lay the foundation for the main topics that would form the basis for new diagnostic criteria and subsequently a proposal to develop a terminology that would incorporate the clinical features of these syndromes while at the same time avoiding any context that might suggest a mechanistic basis. The day was divided into four sessions each of which allowed for the presentation of specific topics that were later debated at times that were considered opportune by the discussants. Because there were no hierarchical differences in the overall

"Ulli" Gebershagen

subject material under review, all five discussants remained throughout the day's sessions. Where differences and debate occurred, there was adequate compliance within the workshop format to allow most to be resolved at the time—a somewhat unusual symbiosis from such a motley group. It was almost as though a certain homage was paid to the long record of historical events that had preceded this workshop and the primitive nature of its medical care at the time. Perhaps the best adjective to describe the success of day one, was `astonishing', the creation of a repository of ideas that would form the basis for the second day, the crux of the workshop! (Table 1.2).

1.3.1.1 Section 1

Clinical Characteristics
Reflex sympathetic dystrophy, clinical features, **Stephen H Butler**
Reflex sympathetic dystrophy, incidence and epidemiology, **Stephen E Abram**
Chronic pain mechanisms, **Terence M. Murphy**
Sympathetically maintained pain, **Peter R Wilson**
Psychosomatic aspects of reflex sympathetic dystrophy, **UT Egle, SO Hoffman**

Perhaps as would be expected, most of the discussion was concerned with the imprecision of the various diagnostic criteria and the hidden or ingenious terms that are used to describe the signs and symptoms of RSD patients. The question of a diagnosis of RSD in the absence of any initiating event and the fact that some of the stigmata of RSD may be present without pain while somewhat of a conundrum, may in fact have CNS origins, a fact of which Wilfrid Jänig frequently reminded the group. Just as is the case in many other medical diagnoses, incomplete diagnostic outliers still justify their inclusion in any future terminology. Of note is the fact that

Table 1.2 AGENDA

Section 1	Section 2	Section 3	Section 4
General aspects of RSD and Causalgia	Basic Research and Pathophysiology	Therapeutic Measures	Recent and Future Techniques
Stanton-Hicks	Robert Boas	Stanton-Hicks	Gabor Racz
RSD Clinical Aspects	Pathophysiology	Sympathetic Nerve Blocks Terrence	Multidisciplinary Care
Stephen H Butler	Wilfrid Jänig	Murphy	Prithvi Raj
Incidence and Epidemiology	SMP and Spinal Hyperexcitability	Neurosurgery for RSD	3-phase Bone Scanning
Steven E Abram	William Roberts	Ronald R. Tasker	Steinert
Chronic Pain Mechanisms	Neuropharmacological Aspects	Peripheral Nerve Stimulation	A Role for Intravenous Clonidine
Terrence Murphy	Ilmar Jurna	Gabor Racz	Christopher Glynn
SMP Diagnosis and Therapy	Neurophysiological Observations	Psychological Support	Intravenous regional anesthesia
Peter Wilson	Erik Torebjörk	David Haddox	Hannington-Kiff
Psychosomatic Aspects of RSD	Blood Flow Dysregulation	Non-invasive Treatment	Discussion and review of the RSD diagnostic criteria
UT Egle	Helmut Blumberg	Edmund Charlton	Boas and Stanton-Hicks

The day was divided into four sessions that were designed to provide the framework for discussions during the second day destined to fulfil the goal of the meeting

whenever a discussion of contemporary criteria takes place the number of patients with a putative diagnosis of RSD will increase.

Ulrich Egle

It was agreed that sex-related differences will probably have to await the results of longitudinal studies there being none currently available, but the peak incidence of RSD seems to be between the fifth and sixth decades.

Hans Nolte

Blumberg noted that there was an apparent difference between the sympathetic activity on venules and arterioles, i.e., suggesting changing patterns of sympathetic activity on these two anatomic structures. Edema as a consequence may or may not occur. Patients presenting with a diagnosis of RSD can generally be summarized into three groups:

1. Those with autonomic disturbances
2. A second group with motor disturbances and a constant fine tremor
3. A third group with sensory disturbances.

The disease entity however affects the entire distal extremity irrespective of its source. Those signs currently accepted include burning pain, hyperpathia/allodynia, temperature/color changes, edema, and hair/nail growth changes.

A sense developed during the proceedings that a sympathetic block, while being an adjunct for other diagnostic criteria might be more useful as an initial instrument to reduce the overall sympathetic influence in the region and at the same time provide analgesia for physiotherapeutic maneuvers.

David Haddox emphasized the fact that there may be a reactive psychosomatic component between the development of RSD, sympathetically maintained pain and the fundamental psychological personality of the patient. Some such patients are often considered to have type-A personalities. All agreed however that there was little literature at the time to support these opinions.

Albert Van Steenberg

1.3.1.2 Section 2

Basic Research into the Pathophysiology of RSD
Pathophysiology of RSD, **Wilfrid Jänig**
WDR neurons and RSD Spinal hyperexcitability, **William Roberts**
Neuropharmacological aspects of RSD clinical and basic physiological observations relating to pathophysiological mechanisms of RSD, **Erik Torebjörk**
Mechanisms and role of peripheral blood flow dysregulation in pain sensation and edema in RSD, **Helmut Blumberg, H, Griesser, M, Hornyak**

Initial discussion focused on the causes of edema in RSD. The underlying imbalance between pre and post capillary pressures need not necessarily result from differences between sympathetic influence on arterioles and venules. It could also be caused by a differential sensitivity of autonomic activity on arterial and venous vascular muscle. Other intrinsic processes and hypoxia at the level of the vessel wall, e.g., the endothelin/NO ratio may be disturbed. In addition, vasoactive substances released from nociceptors can intensify vascular permeability. However sympathetic dysfunction is still present and can be demonstrated by abnormal thermoregulatory responses in the affected areas.

While tissue pressure as a cause of edema may cause spontaneous pain in bone and deep solid tissues like the palms or soles of the feet where tissue expansion is restricted, and pressure can be high enough to stimulate nociceptors, pain may be alleviated by reducing hydrostatic pressure, i.e., lifting the hand or foot above the body. With sensitized nociceptors even small pressure changes may be sufficient to cause pain.

Spontaneous pain is not merely caused by nociceptor activity; in conjunction with allodynia, it can result from the abnormal processing of information from low threshold mechanoreceptors (LTM) or thermoreceptors on convergent or wide dynamic range (WDR) neurons. Further evidence for an underlying central dysfunction is available from reaction studies in patients with mechanical and/or thermal allodynia. Reaction times to innocuous tactile, cold, and warm pulses on the abnormal extremity were modality specific and did not differ from those on the normal size. This would suggest that afferent impulses which elicit pain, originate from low-threshold mechanoreceptors, cold and/or warm receptors respectively. Overall, this session was particularly enlightening and generated new ideas and concepts regarding the pathophysiological processing mediating the syndrome of RSD.

1.3.1.3 Section 3

Therapeutic Techniques in RSD
Sympathetic nerve blocks: their role in sympathetic pain, **Robert Boas**
Intravenous regional sympathetic blocks, **Hannington-Kiff**
Reflex sympathetic dystrophy—neurosurgical approaches, **Ronald Tasker**
Peripheral nerve stimulator implant for treatment of RSD, **Gabor Racz, Royce Lewis, James Heavner, John Scott**
Psychological support of the patient with RSD, **David Haddox**
RSD: non-invasive methods of treatment, **Edmund Charlton**
Multidisciplinary management of RSD, **Prithvi Raj, Jeffrey Cannella, Jennifer Kelly, Karen McConnel, Patricia Lowry**

Technical issues related to the conduct of intravenous blocks, (Intravenous Regional Anesthesia, IVRA) emphasized by Hannington-Kiff dominated the initial discussion. While the use of regional anesthetic blocks of the sympathetic ganglia or somatic nerves of the affected limb was felt appropriate in very severe states, these should only be considered as temporary measures.

"Ron" Ronald Tasker

During the execution of IVRA, distal intravenous injection was felt preferable and safer to reduce the rise in local anesthetic concentration leaking into the systemic circulation. Reiteration of the role of the sympathetic block focused everyone's attention on its perhaps limited capacity to permit reactivation and exercise therapies as the primary purpose of the procedure. The emphasis on sympatholysis as a treatment by itself has probably been the basis for many failed treatments. Neurolytic blocks were not thought to be appropriate in this instance because they tend not to be permanent as was previously thought and while their effect is perhaps long-lasting, because such procedures may be responsible for local scarring around ganglia their repeated use might be impacted by diffusion barriers. This would be a hypothetical consideration that might hinder the successful diffusion (spread) of any future regional anesthetic solutions that are directed at the same targets, however it should be pointed out that past experience in some patients with intractable symptoms have done very well after a neurolytic sympathectomy.

Temperature measurement at digital skin sites was considered a simple, specific, and cost-effective measure for both diagnostic and treatment response assessments.

Sustained and adequate pain relief are two of the most important factors for managing and resolving RSD symptoms (Cooper et al. 1989). Aggressive care, particularly exercise therapy in conjunction with a multimodal approach helps to resolve both physical and negative psychological aspects from which these patients suffer. The overwhelming opinion during this discussion was that if help is needed, most patients will respond to a continuous somatic rather than the more traditional sympathetic blocks before a satisfactory level of pain relief commensurate with their PT can be achieved. The failure of a sympathetic block does not preclude a diagnosis of RSD, notwithstanding its contemporary requirement in support of the diagnosis (Boas 1996).

While it is difficult to draw any strong conclusions from a small case series, the psychological profiles, and apparent differences in response of younger patients with a shorter time interval between injury and treatment would suggest that there may be a two-phase presentation. i.e., RSD may be a syndrome which has an acute severe phase followed by a long (chronic) course of failed management during which the tell-tale signs and symptoms of chronic disease emerge. Careful observation during rehabilitation should anticipate evidence of nascent chronic pain behavior. As discussed, the introduction of an early aggressive multidisciplinary treatment program offers better pain relief and rapid improvement in function as a result of which, many patients avoid the onset of chronic pain. We do not share the view of some clinicians who believe that dramatic pain relief and cure for reflex sympathetic dystrophy can always be obtained from any single mode of therapy used in isolation.

1.3.1.4 Section 4

New Techniques
Three phase bone scanning in RSD, **H. Steinert, O. Nickel, K. Hahn**
An investigation of the role of Clonidine in the treatment of RSD, **Christopher Glynn, P. Jones**

Much of the discussion centered around the use of different adrenergic antagonists as agents for intravenous regional block (IVRA). Guanethidine was felt to be superior and give a more complete and sustained relief then beryllium—with fewer side effects—from the 1 to 3 µg/kg doses used. Because responses were slow to develop and initially less complete, the technique of IVRA was thought to have less utility than sympathetic ganglion block as a diagnostic test. Constructive suggestions were offered seeking to derive research priorities and criteria for clinical diagnosis, grading and treatment responses.

A further definition of the syndrome, RSD in physiological terms would help in defining the functional abnormalities that develop. For each of the presentations recent advances have led to new concepts which should have major implications for further study. Two subgroups of speakers were assigned to provide written submissions including the consensus of these discussions.

Klaus Hahn

Guidelines that set the stage for the future come with the development of animal models and their translation to neurobiological research in RSD and its relationship to clinical research is an epilogue to the Schloss Rettershof meeting.

The rhetorical question is: "can experimental—neurobiological research—be useful to the understanding of RSD and its related syndromes; and as a consequence, will diagnosis and therapy be improved?" The answer is unquestionably "yes." But the hypothesis requires qualification and testing. Animal models used for research on pathophysiological mechanisms of RSD do not exist. Research will only be successful when there is a close interaction between "bench" and clinical investigation. Most research to date has obtained its impetus from clinical practice. This does not imply that experimental scientists and clinicians should work shoulder to shoulder, but rather they should communicate frequently with one another in common terms. The experimental scientist should always remain independent as far as experimental design is concerned.

Research integration—a necessary objective

During the conference we accepted the premise that basic experimental research on different levels of integration is required if we are to understand the pathophysiology underlying RSD. See Fig. 1.5. Clearly since research can only move forward if one assumes that pathophysiology can only be envisaged when and if one is at the same time willing to understand that normal neurobiology and psychological processes are indivisible from neurobiology. Throughout the meeting Wilfrid Jänig took pains to emphasize the symbiotic relationship that is necessary between basic science and clinical research/practice if progress in future laboratory research directions and their clinical application is to be realized. Only in this manner will there be improvements in patient care and better outcomes. The mantra of this relationship is itemized below and reflects many years of basic research into the autonomic nervous system and the catholic dimensions of its influence on the human body, by Helmut Blumberg and Wilfrid Jänig. These remarks relating to clinical and laboratory research are tempered by the following statement: "Basic research should always contribute at different structural levels—Jänig, Blumberg, Torebjörk and Zimmermann.

John Loeser and Manfred Zimmermann

1.3.1.5 Talking Points from Previous Section

1. "Pain behavior" models can be developed when experimental research on such animals shows behavioral change under certain controlled conditions—nerve lesions, chronic inflammation, central lesions; and with experimental interventions that can alter the sympathetic tone—preceding and after sympathetic block.

2. Experimental research on animals can be designed to study single neurons: afferent neurons, motor neurons, sympathetic neurons, or dorsal horn neurons, etc. Well-controlled anesthesia is crucial to the success of in vivo animal research. The objective of these endeavors is to determine mechanisms at the level of a single neuron in relation to specific target organs, i.e., blood vessels, sweat glands, skeletal muscle and the central nervous system. Similar related research needs to be repeated on unanesthetized humans in whom recordings are taken from effector organs or microneurographic recording from peripheral nerves, something with which Eric Torebjörk has had decades of experience.

3. Animal research on isolated systems should involve in vitro work the focus of which should concentrate on morphological, biochemical and neurophysiologic changes of both afferent and efferent neurons after their interaction with each specific target organ. How particular responses of neurovascular transmission and the role of afferents in neurovascular transmission to small pre and post capillary blood vessels take place depends on the site at which evolution and mechanisms of coupling between postganglionic and afferent axons in the periphery, and where neural, environmental, and non-neural structures interact.

4. Basic research should be done in keeping with similar clinical and psychological investigations that involve patients. This should include:

   4.1. Quantitative studies of clinical features, their epidemiology, and the incidence of RSD.

   4.2. Quantitative measurements of sensory, motor, autonomic and trophic changes seen in RSD.

   4.3. Prospective randomized studies that evaluate the efficacy of RSD treatment—physiotherapy, sympathetic blocks, psychological and behavioral therapy, pharmacological therapy and any other supportive measures that are used in a time-dependent fashion, in particular those characteristics that are related to sensory, autonomic, motor and trophic changes.

   4.4. Psychologic/psychosomatic aspects of RSD should be the focus of whether any predisposing psychosocial components exist that might induce the development of RSD or secondly, whether the observed psychological and mental alterations actually occur as a consequence of the condition.

   4.5. Clinical research should be the font of ideas that can only be validated by basic research or vice versa. Quantitative clinical research in and of itself should be the source of comparative data that are responsible for creating acceptable standards for diagnosis and treatment of RSD. A practical means of evaluating abnormal function of the sympathetic, motor and sensory systems should ultimately lead to an agreement regarding both grading and severity of RSD: this will require a common taxonomy.

1.4   Second Day

The second day was the crux of the meeting. It was structured to allow working groups address the different aspects of RSD and causalgia that had undergone such intensive examination during the previous day. Using a modified Dahlem-type format that incorporated detailed instructions to each selected breakout group which in turn would dissect out the minutiae and return to the main group with their final consensus. *(Dahlem Workshops, Silke Bernhard est. 1974; German Science Foundation and the Association for the Promotion of Science Research in Germany)*. This allowed their conclusions and recommendations to be presented and discussed by all attendees at specified intervals throughout the day. Peter Wilson reminded the group that a number of scales had already been developed since the 1950s (Betcher and Casten 1995; Poplawski et al. 1983; Kozin 1986), all of which attempted to define diagnostic criteria that might be used to support a diagnosis of RSD including a recent publication of which he is also an author also developed a diagnostic scale, that included both clinical as well as supportive laboratory measurements, the final number of which would suggest the likelihood of RSD or not (Gibbons and Wilson 1992) (Table 1.3). An outline of proposed diagnostic criteria and the consideration of a new terminology having no mechanistic connotation but would describe in common clinical terms the existing entities, RSD and Causalgia would be the day's goal. Furthermore, the final document should conform to the description of chronic pain entities in the IASP publication, Classification of Chronic Pain: Descriptions of Chronic Pain Syndromes and Definitions of Pain Terms. Interaction during the meeting was intense and difficult with multiple opinions from this disparate group, a credit to the patience of our discussants, much of its progress was kept on the rails and a surprising consensus was actually achieved by the conclusion of the meeting, 1½ days later, before everyone set off for the four corners of the globe.

Table 1.3 These observations discussed during the foregoing session were subsequently published by Gibbons and Wilson in 1992. The clinical signs and laboratory results form a scale which would improve a putative diagnosis of RSD

- Allodynia
- Burning pain
- Edema
- Color changes/hair growth changes
- Sweating changes
- Temperature changes
- Radiographic demineralization
- Bone scan consistent with RSD
- Q-SART or response to sympathetic block

Scoring: 0 = absent criterion; ½. = equivocal; 1. = criterion present

0–2½	RSD absent
3–4.5	Possible RSD
5–9	Probable RSD

Hermann Kreuscher

1.5 Third Day

1.5.1 Loose Ends and Wrap-Up

By the time the group left on day 3, a draft of the proposed new diagnostic criteria that would be sent to the IASP was assembled. An additional request for their consideration was the formation of a Special Interest Group (SIG), Pain and the Sympathetic Nervous System. It is prophetic to note that, by the time the Budapest consensus conference was held in 2003, only minor tweaks in the diagnostic criteria resulting from the passage of time and its validation changed the basic tenor of the draft that was originally sent to the Task Force on Taxonomy of the IASP on conclusion of this 1988 inaugural workshop. Peter Wilson who had worked with colleagues developed a scoring system to diagnose RSD is shown below.

Peter Wilson

The following criteria discussed during the meeting are data from this study in which interscalene block was used for upper extremity pain after failing to respond to a sympathetic block. Q-SART testing. No weight was given to the nine criteria (Gibbons and Wilson 1992: 13; 50) (Table 1.4).

These results lead to more refined putative criteria for Reflex Sympathetic Dystrophy below (Table 1.5):

Table 1.4 Clinical criteria for Reflex Sympathetic Dystrophy

Clinical criteria	
• Allodynia	
• Burning pain	
• Edema	
• Color changes/hair growth changes	
• Sweating changes	
• Temperature changes	
• Radiographic demineralization	
• Bone scan consistent with RSD	
• Q-SART or response to sympathetic block	
Scores:	
0–2½	RSD absent
3–4.5	Possible RSD
5–9	Probable RSD

Scoring: 0 = absent; ½ = equivocal; 1 = criterion present

Table 1.5 Reflex sympathetic dystrophy criteria

Clinical features:	
• Allodynia	
• Temperature/color changes	
• Edema	
• Skin, hair, nail growth changes	
Laboratory tests:	
• Thermometry/thermography	
• Bone X-ray	
• 3-phase bone scan	
• Quantitative sweat test	
• Response to sympathetic blockade	
Score:	
>6	Probable RSD
3–5	Possible RSD
<3	Unlikely RSD

Much of the afternoon after a refreshment break was given to a final editing of the diagnostic criteria that would be submitted to the IASP Task Force on Taxonomy. The dialogue was closed only after final agreement was reached by all in attendance. As is always the case when trying to bring together all the various aspects that were discussed during the previous 2 days, the focus of the discussants during this time was to clarify their directions to each of the participants who would work on each of those loose ends that encapsulate the main purpose of the workshop agenda. It was the intention that each individual or groups would then communicate with one another during the succeeding 2 weeks to generate written responses to the discussants and although working and living in Mainz, Germany, I undertook the responsibility to facilitate the timely transmission of the new diagnostic criteria to the IASP Subcommittee on Taxonomy. After much debate, it was generally agreed that these transcripts (Proceedings) would form the skeleton for a written account that would be printed as suggested by Prithvi Raj, who as Series Editor for Pain

Prithvi Raj

Series published by Kluwer Academic Publishers, Boston, Dordrecht, London, would provide assistance. The book was subsequently published—"Reflex Sympathetic Dystrophy," (Eds) Michael Stanton-Hicks, Wilfrid Jänig, Robert Boas: 1990 Kluwer Academic Publishers, Boston, Dordrecht, London.

1.6   RSD Consensus Criteria

The following definition of RSD represents the final consensus of many views that were expressed by all who attended the workshop at Schloss Rettershof. The final draft as it appears here Is a consensus of these views created by those below named. It will be submitted in this form to the subcommittee on taxonomy of the International Association for the study of pain for consideration as being a clinically more precise instrument than that which presently appears in Supplement 3, 1986 of Pain.

Authors:	
S.E. Abram, M.D.	G.B. Racz, M.D.
H. Blumberg Dr. med.	P.P. Raj, MB; BS
R.A. Boas, MB; BS.	W.J. Roberts, PhD
J.D. Haddox, DDS.	M. Stanton-Hicks, MB; BS
W. Jänig, Dr. med, PhD.	M. Zimmermann, Dr. med
H. Kreuscher, Dr. med, PhD	

Definition
A syndrome of continuous diffuse limb pain, often burning in nature and usually consequent to injury or noxious stimulus, and disuse, presenting with variable sensory, motor, autonomic and trophic changes; causalgia represents a specific presentation of RSD associated with peripheral nerve injury. These features manifest diffusely but not necessarily uniformly in the entire distal extremity. They occur at a variable time after the onset of the syndrome. Spreading approximately and occasionally to the opposite side as the syndrome progresses and are expressed in a "glove and stocking "distribution. Dominating symptoms are spontaneous pain, swelling and weakness.

1.6.1   Clinical Features

Autonomic deregulation. Alterations in blood flow; hypo/hyperhidrosis, edema
Sensory abnormalities. Hypo or hyperaesthesia, allodynia to cold and mechanical
   stimulation
Motor dysfunction. Weakness, tremor, joint stiffness
Trophic changes. Skin, hair, nails
Psychologic reactive disturbances. Anxiety, depression, hopelessness (vis a vie
chronic pain patients)

1.6.2   Diagnostic Tests

Autonomic: bilateral symmetrical multi digital temperature measures by surface
   thermistors or thermography, show consistent discrepancies- cooler or warmer
   on the affected side.
Sensory: load threshold to pin prick, light touch and cold
Motor: reduced measures of strength as well as active and passive range of
   motion (ROM)
Sympathetic block response: should also raise the skin temperature to 35–36 °C
   And abolish the vasoconstrictor response to cold stimulation
Bone scan: three phase bone scanning shows distinctive, diffuse patterns of
   increased flow, pooling and delay.
Nonspecific confirmatory tests: can be used to further quantify changes and follow
   progress These include radiographic densitometry testing for osteopenia, water
   displacement plethysmography for measures of swelling plus tests of sudomotor
   function such as skin conductivity or potentials or quantitative sweat testing.
Staging: cases presented with qualitative differences in pain intensity and clinical
   features, that are not necessarily time or stimulus intensity dependent. Patients
   with the syndrome should be graded according to the intensity of their presenting
   features whether mild, moderate or severe in each case of the categories of sensory, autonomic and motor changes (Bonica 1990). Eponymous or causative designations provide no quantitative or therapeutic benefit for patients with
   sympathetically maintained pain. While some cases may respond with allevia-
   tion of their symptoms by sympathetic block, but still retain a localised pain that
   has one or two hallmarks distinctive of RSD. Staging is not recommended as
   having any value for diagnostic purposes (Fig. 1.4).

1.6.3   Conclusion

The foregoing definition and diagnostic criteria of RSD represents a final consensus
of the many views that were expressed by all who attended the workshop at Schloss
Rettershof.

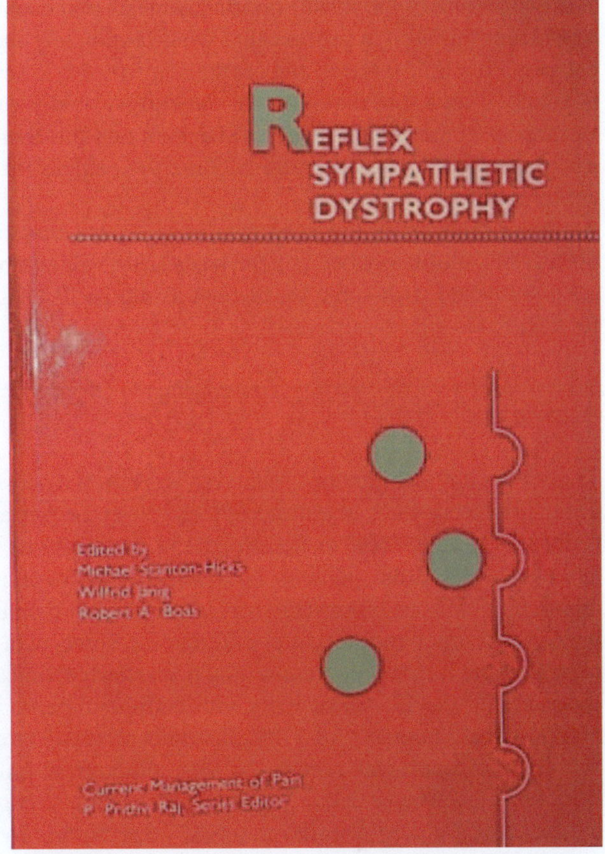

Fig. 1.4 As no Proceedings were recorded at Schloss Rettershof, detailed accounts of each speakers' presentation were assembled for publication in the book, "Reflex Sympathetic Dystrophy", Eds. Michael Stanton-Hicks, Wilfrid Jänig, Robert Boas. Kluwer Academic Publishers, Boston, Dordrecht, London, 1990

Once the proposed diagnostic criteria were published by the IASP Classification of Chronic Diseases, Pain Supplement 1994, work began at each of the subsequent SIG, Pain and the Sympathetic Nervous System meetings that were held at either the annual American Pain Society meeting or an IASP Congress to address the relationship of the sympathetic nervous system to RSD. A steering group which I headed was organized to develop a second workshop in the image of Schloss Rettershof that would address a new Taxonomy based on our current limited understanding of its pathophysiology and the clinical characteristics defined by the contemporary diagnostic criteria. The next such opportunity to organize such a conference was in conjunction with the Annual American Pain Society meeting in Orlando Florida, 1993 (Fig. 2.1).

All contributors were given 6 months to submit their manuscripts. This workshop being the first of what would ultimately become a series of consensus workshops.

Perhaps one of the most important aspects that came out of this workshop quite apart from initiating a process that would ultimately redefine the entities RSD and causalgia in clinical terms and set the stage for basic scientists to explore its pathophysiology and stimulate clinical curiosity would be the acceptance by IASP to form a Special Interest Section (SIG Pain and the Sympathetic Nervous System). This latter development would become the vehicle for many future meetings and conferences.

At the conclusion of the workshop, a statement that summarized the 2½ days of discussions ultimately became the basis, after validation, for what has subsequently been published with some refinement as the IASP Diagnostic "Budapest" Criteria for CRPS.

Appendix

Transcripts and publications related to workshop discussions.

In 1988 very little about the epidemiology or prevalence of RSD was known. Carron and Weller (1974) described 158 patients who met rigid criteria for RSD amongst 1156 clinic patients (10.7%). The ratio of male to female prevalence reported by various observers seemed to reflect the nature of the population rather than any distinct male/female differences. Some data were available to discuss regarding the incidence of causalgia, particularly Bonica 1953 who cited some 296 cases among the 12,335 patients with peripheral nerve injuries with which he was associated, an incidence of 2.4%. Echlin et al. (1949) reported that during the Civil War the incidence was reportedly 38% although in only 2% of his series did the symptoms last longer than a few days. Also acknowledged during the meeting was the fact that children also develop RSD. See reports by Bernstein (1978), Stanton et al. (1993), Goldsmith et al. (1989) and Kozin et al. (1977).

In his book "Management of Pain," 1953 Bonica published an extensive list of conditions that have been known to trigger the pathological process. Some of these include sprain's, fractures, lacerations, crush injuries, amputations, and burns. Surgery, occupational factors such as repetitive microtrauma and pneumatic tool operation add to the variety of triggers that are known to precede the development of this condition and several visceral states such as myocardial infarction, and some neurological states such as stroke can precipitate the onset of RSD, not infrequently, in this latter case, the upper extremity.

Sympathetic Nervous System

The role of the Sympathetic Nervous System (SNS) was a recurrent theme during the meeting. While not generally a point of discussion in relation to acute, chronic, or neuropathic pain states, there was uniformity of opinion which seemed to suggest that the SNS is significantly involved in RSD pathophysiology. Most of the neurologists and neurophysiologists present, in particular Helmut Blumberg, also suggested that similar activity is also evident in other neuropathic pain states like post herpetic neuralgia, radiculopathy, and certain peripheral neuropathic and central pain syndromes. A number of work-related conditions such as repetitive strain injury, cumulative trauma disorder, overuse syndrome (Musicians, violinists etc.) also fall into this category. The progressive pain of a burning nature throughout the ipsilateral extremity, impairment of function, edema, sensory changes, and autonomic dysfunction characterize these symptoms.

William "Bill" Roberts

Having William (Bill) Roberts, PhD in the group was salutary as he pointed out from his own research, Roberts (1986), that low-threshold mechanoreceptors (LTM) probably mediate sympathetically maintained pain (SMP). Although this has been tested physiologically none of the studies at the time provided support for a hypothesis that sympathetic activation of primary afferent nociceptors is actually responsible for the allodynia manifest in RSD. Several classes of LTM are sympathetically activated. Support for this by Loh et al. (1981) demonstrated that trauma could lead to: (1) sensitization of spinal wide-dynamic range (WDR) neurons, (2) sympathetic activation of LTM can result in excessive firing of hyperexcitable WDR neurons and therefore spontaneous pain and (3) Mechanical activation of LTM results in excessive firing of hyperexcitable WDR neurons and therefore spontaneous pain. The recent studies of patients with SMP, Roberts and Foglesong (1988), Jänig and McLachlan (1987), support the evidence that LTM mediate spontaneous pain and allodynia in RSD (Fig. 2.2). If graduated pressure is applied to a limb in order to progressively block conduction in nerves that innervate the painful area, the sensation of touch is lost first as the largest A-β axons are blocked. Patients with SMP who perceive a pressure block affecting only A-β fibers lose their spontaneous pain and mechanical allodynia (Ochoa 1994; Roberts 1986). There was agreement that SMP is mediated by LTM and that abolition of mechanical allodynia during sympathetic block is most likely explained by loss of facilitation of spinal WDR neurons, i.e., block of sympathetic efferent activity may reduce tonic activity of LTM. It is interesting to recall Claude Bernard's experiments on the sympathetic nervous system of rabbits in which he demonstrated the relationship of pain to sympathetic activity when he interrupted this pain behavior by blocking the sympathetic supply to the region that he was studying.

Also, should be mentioned is that during this meeting the syndrome was still being discussed in terms of the dreadful stages, namely 3:

1. Acute hyperemia,
2. Dystrophic ischemia.
3. Chronic atrophia,

interestingly, 3 degrees of severity were assigned to the clinical presentation

(a) Mild
(b) Moderate
(c) Severe.

Obviously, there would be no single test or group of tests that could be either specific or even reliable. The diagnosis until such time as a mechanism is discovered will remain a clinical experience and gestalt.

All present acknowledged that the current IASP taxonomy is inadequate, requires redefinition using the characteristics shown below.

Helmut Blumberg

Psychosomatic Aspects

The psychosomatic aspects of RSD were discussed at length during the early part of the meeting. Many reports describe such patients as anxious and tense, exhibiting neuroautonomic stigmata. Langston and Covan described a premorbid personality with pronounced characteristics of hypertension and a tense nervous system (Langston and Cowan 1955). The term Sudeck-personality was described by Hübner in 1957. The behavioral scientists present at the meeting were quick to point out that the only pathogenic concept, and one that is explicitly psychogenic was developed by de Takats in 1943. They considered the patients' psychological status to be of vital importance to expression of autonomic reflexes and therefore to the subsequent development of RSD. While initially accepting that these earlier observations of RSD point to a group of psychosomatic disorders whose principle pathogenic aspect is determined by premorbid psychological elements and somatic symptoms, the entire group and a very loquacious David Haddox felt that the reliability and subsequent validation of such views had not been confirmed. Furthermore, such psychological disorders are not necessarily a cause but rather may result from the experience of chronic pain. Engel (1959) who Introduced the concept of Psychodynamics in chronic pain, became a major point of discussion since the social relationships are of major consequence affecting both disposition of the disease and its subsequent clinical course. At the time, only meager data were available to coincide with our observations of behavioral changes in chronic pain patients: the influence of a patient's relatives is of similar importance to RSD patients—affecting the duration of the condition, particularly during family conflicts. Such patients are susceptible to a basic narcissistic dynamic. Blumer and Heilbronn (1982) already suggested that chronic pain is like a variant of depressive disease, often colored by inertia and self-indulgence or unconsciously accusatory behavior: "look what you have done to me, now see what you get," in a way similar to hypochondriacal patients which may take on a more depressive or hysterical course reflecting somewhat their own basic personality. Egle and Hoffman, both presented unpublished data of their recent eight patients who had tried to initiate confrontation between different treating specialties, four of which discontinued their treatment after having aggressive arguments with their treating physicians. Such reactions confirm the extreme susceptibility and a basic narcissistic dynamic in such RSD patients, nevertheless, one can still accept the fact that such patients represent a relatively small percentage of all RSD patients.

Pathophysiology

Wilfrid Jänig postulated that the following pathophysiological processes are very likely to be a component in the genesis of RSD, shown below (Fig. 1.5).

1. **Primary efferent neuron**, postganglionic sympathetic neuron Retrograde cell reactions: biochemical changes as a reflection of repair involving neurons in their entirety; A decrease in neuropeptide content of primary afferent neurons; incipient atrophy and degeneration of neurons Anterograde cell reactions: A mismatch of axon-receptor and postganglionic-target organ fibers; subsequent neuroma formation.

   The membrane properties of afferent sprouts develop abnormal resting activity, abnormal chemosensitivity, e.g., nor adrenaline and mechanosensitivity of sprouting afferent neurons.

   Changes in the receptive properties of intact primary afferent fibers due to changes of micromilieu—ischemia, release of inflammatory mediators and vasoactive substances; sensitization of efferent receptors.

   The development of ephaptic transmission among afferent fibers in the lesioned territory and/or nerve, could be a possible amplifier for afferent activity.

   The development of "crosstalk" between postganglionic sympathetic fibers and afferent fibers.

   Changes or failure of synaptic transmission from preganglionic to postganglionic neurons in sympathetic ganglia and from primary afferent neurons on dorsal horn neurons.

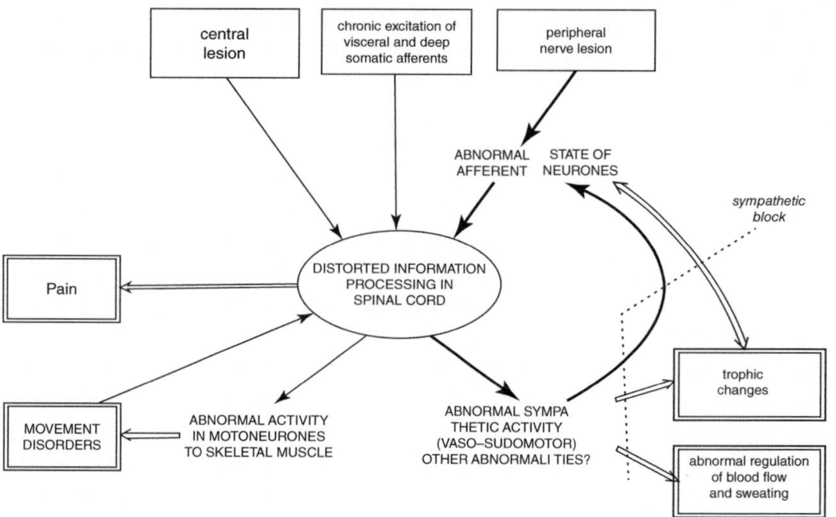

Fig. 1.5 Hypothesis: General hypothesis about the neural mechanisms that can generate reflex sympathetic dystrophy. Following peripheral nerve lesions, central lesions, or chronic stimulation of visceral and deep somatic afferents a Circulus vitiosis (Black arrows) ensues. Important to this concept is the excitatory influence of postganglionic, sympathetic axons on primary afferents In the periphery. This influence leads to orthodromic afferent impulse activity. It can however induce anti-dromic Impulses in unmyelinated afferents in the periphery, thereby in turn, contribute to trophic changes by releasing of substances in the axon terminals. The consequences are vasodilatation and plasma extravasation. Such trophic changes may also influence coupling between postganglionic axons and afferent axons (interrupted double arrow), modified from Blumberg and Jänig (1983), Jänig and Kollmann (1984). The figure modified from Figure 3 taken from the article "The puzzle of "Reflex Sympathetic Dystrophy"": Mechanisms, Hypotheses, Open Questions by W. Jänig, Reflex Sympathetic Dystrophy: A Reappraisal. Progress in Pain Research and Management. Vol 6. Eds. W Jänig and M Stanton-Hicks. IASP Press, Seattle. 1996. With permission

2. **Spinal cord**

Changes in the distribution of primary afferent axon terminals in the spinal cord—degeneration, sprouting.

Unmasking of synapses from primary afferents at the second order neurons; leading to the formation of new aberrant synaptic connections.

Biochemical changes in dorsal horn neurons—with a decrease of peptide content and initiation of supersensitivity of neurotransmitters and modulators due to deafferentation.

Change of inhibitory control mechanisms in the dorsal horn—both local and descending from the brainstem; change of receptive fields and discharge proper-ties of dorsal horn; change of somatotopic map.

Development of abnormal discharge properties and abnormal reflexes in sym-pathetic neurons supplying the affected extremity—skin, deep somatic tissue etc.

Development of abnormal discharge properties in motor neurons subserving reflexes in the affected extremity.

3. **Target organs**
   1. Adaptive supersensitivity and sub-sensitivity of autonomic effector organs on degeneration and regeneration of their neural supply occurs; these influences are responsible for abnormal responses to neural stimulation, the release of local circulating substances and abnormal reaction to external influences, e.g., changes in environmental temperature
   2. The abnormal regulation of blood flow through the skin and subcutaneous tissues
   3. A mismatch of vasoconstrictor activity to different sections of vascular beds, together with an increase of filtration pressure in capillary beds
   4. An increase of filtration pressure in capillary beds the subsequent chronic formation of edema due to imbalance between pre and post capillary vasoconstrictor activity; Increase in vascular permeability due to release of vasoactive substances, e.g., from primary afferents or other activity in local cells
   5. Atrophy and degeneration of dermis, capillaries, subcutis, joint capsules; mineral loss in bones

All five levels of integration interact reciprocally by neural and biochemical signals. Changes in the foregoing can perturb the general homeostasis and may in turn lead to those physical phenomena ascribed to RSD. A lot of unrelated events can induce the same clinical phenomena: trauma with or without nerve lesions, visceral events—e.g., coronary complications, events in the deep somatic domain, e.g., skeletal muscle, tendons, fascia, bone etc., and central lesions. It is therefore theoretically possible for clinical phenomena of RSD to be initiated from the neocortex/limbic system without any evidence of trauma. Why should trivial trauma result in the generation of RSD? The spinal cord, and in particular the dorsal horn with this its neuronal systems—cutaneous, deep, somatic, visceral; from spinal centers and its output systems—sympathetic, skeletomotor, brainstem to thalamus are all components of normal neurological function. As such, RSD is a neurological disease and the sympathetic outflow is but one, although important component in the pathology of the syndrome. It is therefore never advisable to simply focus on the sympathetic nervous system as necessarily being at the crux of the matter.

Wen Hsein Wu

Interest in the sympathetic nervous system (SNS) during the past few years has not only increased exponentially but it has been responsible for sparking a resurgence of laboratory activity to answer the question, *"in what way and how might the SNS be involved in this activity"?* (Basbaum and Besson 1991; Fields and Basbaum 1999; Jänig and Schmidt 1992). This has had the salutary effect of dispelling many of the previously held controversies and has also laid to rest the idea that the SNS is the primary instigator of the clinical entities RSD and Causalgia. Without understanding the mechanistic basis by which the SNS interacts with the neuropathic, nociceptive and inflammatory processes already identified in the pathophysiology of the syndrome, there is little to question why sympathetic blocks have been successfully used over the preceding decades to manage pain in a comparatively large number of patients. Certainly, the focus for the Orlando workshop was to address this issue which amongst all the other topics that would be discussed to provide a redefinition of RSD having only an indirect but most likely, multiple interactions with the SNS.

I decided on the venue for this meeting after having several discussions with members at SIG, Pain and the Sympathetic Nervous System meetings and with the suggestions and logistic help that I obtained from Louisa Jones, Executive Officer, IASP. The meeting site was coordinated with the Organizers of the American Pain Society whose Annual Meeting was to be held in Orlando. We selected a nearby small self-contained conference centre just adequate to accommodate our small group for 2 days in the centre of a large local golf course. As such, it provided all the amenities, and audio-visual services on-site for the Workshop and being as they say, "off the beaten track" it provided the isolation that was necessary for the conduct of our mission. As discussed below, the consensus at this Workshop was

© The Author(s), under exclusive license to Springer Nature Switzerland AG 2024
M. Stanton-Hicks, *The Evolution of Complex Regional Pain Syndrome*,
https://doi.org/10.1007/978-3-031-54900-7_2

reached with a final agreement that the term, RSD would be replaced by Complex Regional Pain Syndrome (CRPS). All agreed, including Harold Merskey, who as co-editor of the Classification of Chronic Pain: Descriptions of Chronic Pain Syndromes and Definitions of Pain Terms, IASP Press, acquiesced stating that the term RSD should be identical with the new term, CRPS Type 1 and Causalgia would be identical with CRPS Type 2. It was universally accepted for practical as well as historical reasons that Mitchell's description and term would for now be retained within the new terminology, whether or not a common denominator for CRPS I and CRPS II is subsequently found. Furthermore, the term Sympathetically Maintained Pain (SMP) describes what is a symptom only and should not be used to define any type of pain syndrome. The definitions of CRPS I and II are now described in the Second Edition of Classification of Chronic Pain (Merskey and Bogduk 1994; Stanton-Hicks et al. 1995). When one is considering any changes to a taxonomy or even creating a new terminology,

It is worth paraphrasing John Bonica (1994), who in the Introduction to the Classification of Chronic Pain stated, *"The development and widespread adoption of universally accepted definition of terms and a classification of pain syndromes among the most important objectives and responsibilities of the IASP. It is possible to define terms and develop a classification of pain syndromes which are acceptable to many, or be it not at all, readers and workers in the field; even if the adopted definitions and classifications are not perfect they are better than the Tower of Babel conditions that currently exists; adoption of such classification does not mean that it is fixed for all time and cannot be modified as we acquire new knowledge; and, the adoption of such a taxonomy with the condition that it can be modified will encourage its use widely by those who may disagree with some part of the classification. This in fact has been the experience and chronology of such widely accepted classifications as those pertaining to heart disease, hypertension, diabetes, toxaemia or pregnancy, psychiatric disorders, and a host of others. I hope therefore that all IASP members will cooperate and use the classification of pain syndromes after this is adopted by the IASP to improve our communications systems. This will require that they be incorporated in the spoken and written transfer of information, particularly scientific papers, books etc., and in the development of research protocols, medical records, and databanks for the storage and retrieval of research and clinical data."*

While transcripts are not included in this description, the following topics made up the body of the workshop discourse which followed the previously used Dahlem-type format (Table 2.1).

Table 2.1 Participants and their affiliations attending the Orlando Consensus Workshop

Marshall D Bedder MD Advanced Pain Management Group, Portland OR USA

Gary J Bennett PhD Neurobiology and Anesthesiology Branch, National Institute of Dental Research, National Institutes of Health, Bethesda MD USA

Helmut Blumberg Dr. Med PhD Neurochirurgische Universitäts Klinik, Klinikum der Albert-Ludwigs-Universität, Freiburg, Germany

Robert A Boas MBCHB, FANZCA, FRCA pain clinic, Auckland Hospital, Auckland, New Zealand

Edward C Covington MD Chronic Pain Rehabilitation Program, the Cleveland Clinic Foundation, Cleveland, Ohio, USA

Wilfrid Jänig Dr. med, PhD Physiologisches Institute, Christian-Albrecht's-Universität zu Kiel, Kiel, Germany

Martin Koltzenburg Dr. med Department of Neurology, University of Würzburg, Würzburg, Germany

Phillip A Low MD, PhD Autonomic Disorders Research Center, Department of Neurology, Mayo Clinic, Rochester, Minnesota, USA

Gabor B Racz MD Department of Anesthesiology, Texas Tech University Health Science Center, Lubbock, Texas, USA

P Prithvi Raj MD Pain Medicine Center, Los Angeles, California, USA

Richard L Rauck, MD Department of Anesthesiology, Bowman Gray School of Medicine, Winston-Salem, North Carolina, USA

William J Roberts PhD R.S. Dow Neurological Sciences institute, Portland, OR, USA

Paolo Sandroni MD Autonomic Disorders Research Center, Department of Neurology, Mayo Clinic, Rochester, Minnesota, USA

Michael Stanton-Hicks MB BS Dr. med, FRCA, ABPM Pain Management Center, the Cleveland Clinic Foundation, Cleveland, Ohio, USA

Catherine L Willner MD Autonomic Disorders Research Center, Department of Neurology, Mayo Clinic, Rochester, Minnesota, USA

Peter R Wilson, MB; BS, PhD Pain Clinic, Mayo Clinic, Rochester, Minnesota, USA

Session 1: Opening and Program

Introduction and goals from Schloss Rettershof. **Michael Stanton-Hicks MB; BS, Dr med,**

Mechanisms, Questions, Hypotheses—**Wilfrid Jänig, Dr med, PhD,**

Afferent Pain Mechanisms, Neuralgia—**Martin Koltzenburg, Dr med,**

Animal Models—Clinical Contribution—**Gary Bennett, PhD, William Roberts, PhD,**

Quantitative Sensory Testing (QST), **Donald Price PhD**, Gracely PhD, (absent), Bennett, PhD

Discussion and Breakout Groups

Psychological aspects and Placebo Problem **Ed Covington, MD. David Haddox, DDS, MD**
Clinical aspects of CRPS/RSD **Philip Low, MD, Peter Wilson MB; BS, PhD**
SMP **Price, PhD,** Gracely PhD (absent)

Discussion and breakout groups

General plenary session and breakout presentations
Session 2: Introduction and Program. Michael Stanton-Hicks, MB; BS
SMP and relationship to CRPS/RSD **James Campbell, MD**
Complex Regional Pain Syndromes **Robert Boas, MB; BS**
Reflex Sympathetic Dystrophy in Children, **Robert Wilder, MD, PhD**
Diagnostic Algorithm **Peter Wilson, MB; BS, PhD**, Marshall Bedder, MD, Richard Rauck, MD
Regional Anesthetic Procedures for Diagnosis, **Prithvi Raj,** MD, Michael Stanton-Hicks, MB; BS, Dr. med, Gabor Racz, MD, Samuel Hassenbusch, MD, PhD

Discussion and Breakout Groups

Development of a Taxonomy, **Harold Merskey, DM**
Summary and Epilogue Michael Stanton-Hicks, MB; BS, **Wilfrid Jänig, Dr med, PhD**

Most of the afternoon session was devoted to presentations and discussions around the new clinical description for CRPS, its diagnosis and final preparations for the next Consensus Workshop. All members were appraised of the regular SIG, Pain and the Sympathetic Nervous System at which sessions in conjunction with APS and IASP calendar of scheduled meetings will follow any new progress concerning the terminology and diagnostic criteria (Participants are shown in Fig. 2.1).

Because of its importance to the outcome of the meeting the sympathetic nervous system (SNS) was a leitmotiv orchestrated by Wilfrid Jänig throughout the meeting. As the main regulator of body function (homeostasis) in most tissues and as the dominant efferent component of the central nervous system and peripheral nervous system, the SNS both *centrally* and *peripherally* controls blood flow within organs—cardiovascular system; digestive system; visceral systems; reproduction and thermoregulation.

The neuroeffector connections via the *postganglionic ganglia* to their specific targets; blood vessels, glands, enteric neurons, cardiac muscles, non-vascular smooth muscle; *paravertebral ganglia* are merely *relay* stations although some do have the capacity to interact and integrate *afferent* input from viscera (Jänig 1985; Jänig 1995; Jänig and McLachlan 1987). The relationship of the SNS to pain is twofold, i.e., (1) a generalized or specific localized response to noxious stimuli vis a vie, trauma and (2) tissue damage without a nerve lesion can cause spontaneous pain with both burning and hyperalgesic components (Jänig 1995). This type of pain can often be relieved by a sympathetic block. The confusion that exists since RSD as a syndrome was first described (Bonica 1997), implied a mechanistic association with the SNS which during the course of the workshop underwent intensive scrutiny and was dispelled as such by acknowledging our improved understanding of its clinical phenomenology. A

Fig. 2.1 Participants at the Orlando workshop
  Front Row: Robert Wilder, Catherine Wilner, Donald price, Nagy Mekhail, Ed Covington
  Second Row: Ralf Baron, Wen-hsein Wu, Gabor Racz, Phillip Low, Michael Stanton-Hicks, Peter Wilson, Harold Merskey, Sam Hassenbusch
  Back row: Bill Roberts, Marshall Bedder, Richard Rauck, David Haddox, Helmut Blumberg, Robert Boas, James Campbell, Wilfrid Jänig, Martin Koltzenburg

consensus statement that would encapsulate this relationship in purely clinical terms was finally reached after many colourful proposals had been made—Robert Boas, "Complex Regional Autonomic Pain Syndrome" (CRAPS)—the term "Complex Regional Pain Syndrome" (CRPS) was eventually chosen (Jänig et al. 1991; Merskey and Bogduk 1994; Stanton-Hicks et al. 1995). This umbrella term would include regional pain and other sensory changes succeeding a noxious event together with the classical changes of skin temperature, color, abnormal sweating, edema and not infrequently, motor/trophic abnormalities. Two types, CRPS 1 which corresponds to RSD and CRPS 2, i.e., associated with a known nerve lesion—"Mitchell's Causalgia"—are described. Deliberations concerning how SMP should be categorized were arduous, the final deliberation being to relegate it to that of a symptom status and to acknowledge that it is also frequently identified as a component of other pain syndromes.

The symptom characteristics of CRPS constitute items below (Blumberg 1988; Blumberg and Jänig 1994).

- **Pain**—spontaneous, hyperalgesia, allodynia
- **Abnormal regulation**—of blood flow and sweating
- **Edema** of skin and subcutaneous tissues
- **Trophic** changes of the skin, appendages and subcutaneous tissues
- **Different symptoms** are variable in their expression and combination
- **Active and passive** movement disorders and physiologic tremor

The variability and *expression* of clinical phenomena is highlighted by an increasing bibliography of psychophysical measurements on patients with chronic CRPS (Ochoa 1992; Lynch 1992; Verdugo and Ochoa 1994; Price et al. 1992; Arner 1991). For example, the constant burning pain, touch-evoked allodynia and cold allodynia all of which can be dependent on sympathetic innervation whereas numbness, dysesthesia and paroxysmal pain; heat-evoked hyperalgesia and hypoalgesia may be independent of sympathetic activity. As the participants found this aspect of the discussion 'heavy sledding', Helmut Blumberg suggested the following set of symptoms would help to clarify the many opinions and differential diagnosis of CRPS types I and II (Table 2.2).

Separation of CRPS type II from a diagnosis of CRPS I acknowledged some specific differences in the triad of sensory, autonomic and motor symptoms in which the spread of CRPS II beyond the lesioned site is less or even absent in some cases. Changes related to vasodilation and later vasoconstriction.

Recent data were presented in which two types of patients—warm and cold—were subject to whole body cooling and warming using a thermal suit in which the environmental temperature could be standardized (Wasner et al. 2001; Bini et al. 1980). Measurements of the subjects show dynamic change in the vascular responses at the affected side in comparison with the healthy side. Those patients classified as *warm patients* were found on average to have a temperature that was 2–3 degrees warmer than the healthy side (environmental temperature kept at 20 °C) Also noted when the temperature was slowly increased to 4–5 °C, the affected limb became warmer, but the skin temperature increased much more slowly on the affected side than did the temperature increase on the unaffected limb. In comparison, the temperature decreased rapidly to much lower values in the affected limb of *cold patients* than those on the healthy side after total body cooling. The maximum temperature

Table 2.2 The differential diagnosis of CRPS syndromes type I and type II (Blumberg 1988; Blumberg and Jänig 1994; Veldman et al. 1993)

CRPS I	CRPS II	
Etiology	Lesion (any trauma)	Partial nerve lesion
Localization	Distal extremity	Any peripheral body site
Symptom spread	Obligatory	Rare
Spontaneous pain	Common	Predominantly superficial
	Mostly deep & superficial	
	Orthostatic component	No ortho static component
Mechanical allodynia	Especially in patients with symptom spread	Obligatory in nerve territory
Autonomic symptoms	Generalized the distally particularly in spreading cases	Related to nerve lesion
Motor symptoms	Generalized	Related to nerve lesion
Sensory symptoms	Generalized distally spreading cases	Related to nerve lesion

The clinical characteristics of CRPS I and CRPS II, their regional expression and differences when a partial nerve lesion is present or not are compared in the table. Modified from Table 1. Baron, Blumberg and Jänig, Chapter 2. Clinical characteristics of patients with complex regional pain syndrome in Germany with special emphasis on vasomotor function, particularly during the acute stage are most severe in the nerve vicinity. Impairment of motor function may be related to motor neuron damage. Late changes are related to sympathetic impairment and the development of increased sensitivity to , cold temperatures, cathecholamines and the up regulation of adrenoceptors (Fleming and Trendelenburg 1961; Jobling et al, 1992; Jänig and Koltzenburg 1991; Koltzenburg and Torbjörk 1995) Reflex Sympathetic Dystrophy: A reappraisal. Progress in Brain Research and Management, Vol 6 Eds. W Jänig and M Stanton-Hicks. IASP Press, Seattle, 1996 difference during these conditions was 4–5 °C after whole-body warming these side differences were no longer apparent

The following remarks summarize the day's proceedings:

2.1   CRPS1

- CRPS can arise from a multitude of many types of traumata. It is defined by a triad of sensory, autonomic, and motor symptoms that occur distally in the affected extremity most notably they are generalized and independent of any type or location with the preceding injury. These symptoms are not unique to any single peripheral nerve distribution and are disproportionate in their intensity to the inciting event.
- Most cases of CRPS I are acute and should be diagnosed early using the following clinical features:
  1. Spontaneous, deep, diffuse pain in most patients. Both cutaneous and deep hyperalgesia together with other sensory changes are frequently found. Allodynia is present in most cases depending on the relationship in time to the initiating event.
  2. Autonomic changes include edema, disturbed blood flow with temperature side differences and frequently Altered sweating—hyper/hypohidrosis.
  3. Motor changes are obvious. Because of the reduction in range of motion (ROM), reduction of muscle strength and frequently an increase in physiological tremor of the distal extremity are seen.
  4. Trophic changes of the skin, its appendages, muscles, bones, and joint contractures are later consequences of CRPS I
  5. When untreated, CRPS1 may be present for months or years. It can change into an intermittent form but spontaneous remissions do occur.
  6. Underlying these clinical characteristics is increasing evidence that CRPS1 is a *central nervous system* disorder. The vascular abnormalities maybe a reflection of a unilateral failure of thermoregulation in the extremities. Such abnormalities may also result from abnormal reflex responses in vasoconstrictor neurons or to body temperature stimuli on the affected side; similar blood vessel abnormalities are seen in CRPS 2 where the associated nerve lesion maybe due to the peripheral impairment of sympathetic function and denervation.

2.2   CRPS II

CRPS II occurs after a peripheral nerve lesion. The principal symptoms of which are noted at the site of the affected nerve but may not infrequently be demonstrated spreading from the region as spontaneous pain, mechanical and cold allodynia, skin blood flow and sweating abnormalities. These phenomena are generally outside of the injured nerve territory. Functional disturbances may occur. Trophic changes are rare. It is always possible for CRPS II and CRPS I to occur in the same patient.

Participants from the Mayo Autonomic Reflex Laboratory who were in the group had done a side-by-side comparison of sudomotor and vasomotor indices in 12 patients with RSD/SMP, the results of which are shown in Table 2.3.

Table 2.3 Mayo study of 12 patients diagnosed with RSD/SMP (Low et al. 1994)

Case	Age	Pain duration	pain	swelling	VM	X-ray Demin.	Bone scan	Severity	Treatment response
1	51	17	+	+	+	+	?	mild	?
2	26	4	+	++	+	–	ND	mild	Y (PM&R)
3	37	4	+	+	+	–			
4	38	4	++	?	?	–	–	moderate	Y (symp block)
5	73	6	++	+	+	–	–	Moderate	
6	50	10	++	+	++	–	ND	Severe	Y (symp block)
7	26	9	+	+	+	–			
8	19	10	++	+	+	–	ND	Severe	Y (symp block)
9	39	18	+	+	++	ND	Moderate		
10	44	36	++	++	?	–	ND	severe	Y (symp block)
11	24	2	++	++	++–			Severe	Y (symp block)
12	14	12	+	+	?	–	–	Mild	Y (symp block)

The pain duration (yrs), severity, swelling and vasomotor changes are recorded together with the bone scan, x-ray images and response to treatment are shown
PM & R-physical medicine and rehabilitation; symp-sympathetic, ND no data, Y yes
Modified from Table 2, Low et al. Laboratory Findings in Reflex Sympathetic Dystrophy: A Preliminary Report. Clin J Pain. 1994, Raven Press Ltd. New York. with permission

As can be seen in the table, all patients had diffuse pain hyperalgesia and allo-dynia and most patients had vasomotor abnormalities that were diffuse extending beyond the peripheral nerve dermatomes and maximal distally. Skin blood flow abnormalities were found in 44% of patients and sweating abnormalities measured by resting sweat output (RSO) and evoked sweat response (QSART) were measured in two of three (RSO) and three of four (QSART) patients (Low et al. 1983). The group agreed that sweat measurement could be a useful as an aid to diagnosis.

Retrospective data in 407 patients over an almost 10-year period from the same investigators suggested that blood flow measurement was not useful and that tempera-ture measurement from an averaging algorithm suggested a side difference of 1.5 °C on the affected side. These results were superseded by a prospective study by Low et al. (1994), who provided the group with preliminary insights from these data (Table 2.4).

Richard Gracely

Table 2.4 Preliminary diagnostic data (pre-publication) from the Mayo Clinic Autonomic Laboratory

Parameter	Definite	Probable	Possible	Not RSD/SMP

Allodynia
  A.   Touch
  B.   Pressure
  C.   Movement
	3/3	2/3	1/3	0/3

Vasomotor
  A.   History
  B.   Exam
	4/4	$\geq$2/4	$\leq$1/4	0/4

Swelling
  A.   History
  B.   Exam
	ND	ND	ND	ND

The following scores are based on a prospective study using a seven-point scale below were presented by Phillip Low during the meeting

Definite RSD/SMP	7
Probable RSD/SMP	4–6
Possible RSD/SMP	2–3
Not RSD/SMP	0–1

Phillip Low

At this juncture of the session, the attendees considered the recommendation to be the basis for a diagnosis of RSD/SMP and in the absence of any etiopathogenetic support, (1) pain severity, (2) distribution of pain—diffuse, (3) Allodynia are the cardinal characteristics.

On the second day, accounts of RSD/SMP in children became well established at this workshop, particularly the number of recently cited reports that were revealed by Bob Wilder and cited by Kesler et al. (1988), Schiller (1989), Touzet et al. (1991), Wilder et al. (1992), Stanton et al. (1993). While the clinical picture is not dissimilar to that of the adult presentation, the symptoms differ only in an age-related manner, in a perhaps more emotional context and with a much higher preponderance of adolescent females. The treatment has mostly been conservative with physical therapy and TENS, the principal objective being a rapid return to their previous function. Cognitive behavioral measures and biofeedback are particularly appropriate for children (Richlin et al. 1978; Barowsky et al. 1987; Kawano et al. 1989). A few patients may need more invasive treatment with nerve blocks. Sympathetic nerve blocks while not being considered as the first line of treatment, their use to facilitate physical therapy can be salutary. Resolution of symptoms occurred in at least half of these children with some having continuing pain particularly during times of stress and a few having total disability requiring chronic pain management to afford the highest possible quality of life.

"Bob" Wilder

To address the nomenclature and classification of CRPS/RSD, Robert Boas presented the diagnosis and signs from one of the working groups whose main charge was to find out why it is so difficult to achieve any agreement when it comes to diagnostic criteria. The entire group acknowledged the conclusions outlining the diagnostic criteria. These would become the framework for their ultimate incorporation in the terminology.

Bob Wilder postulated the limitations that would need to be addressed if criteria defining pediatric CRPS are met, below:

1. Until pain clinics and trained specialists are available, having a dialogue with those disciplines which have no association with pain management will remain the main impediment to the initiation of appropriate management of children and adolescents who suffer from CRPS/RSD.

2. Nervous system excitation has proven difficult to identify in clinical and laboratory research
3. Treatment based on interruption of sympathetic function has failed to provide lasting benefit for many apparently typical CRPS children
4. Lack of animal models has limited basic research
5. Absence of a diagnostic "gold standard" focused attention on diagnostic procedures such as the thermography, three-phase bone scanning, sweat testing, quantitative sensory testing (QSART) and phentolamine infusions, which have helped promote a more extensive critical examination of the syndrome

Everyone agreed that the previous vicious cycle hypothesis (Livingstone 1943; Bonica 1953, 1979, Bonica 1990), while having some application to the pathophysiology of CRPS/RSD was no longer valid to sustain an RSD terminology or it's therapeutic consequences.

As the basis for a terminology, *Complex* should be a component; *Regional* because of its distribution and *Pain* is the sine qua non because pain is disproportionate and includes spontaneous pain, burning pain and thermal or mechanically induced allodynia (Fig. 2.2).

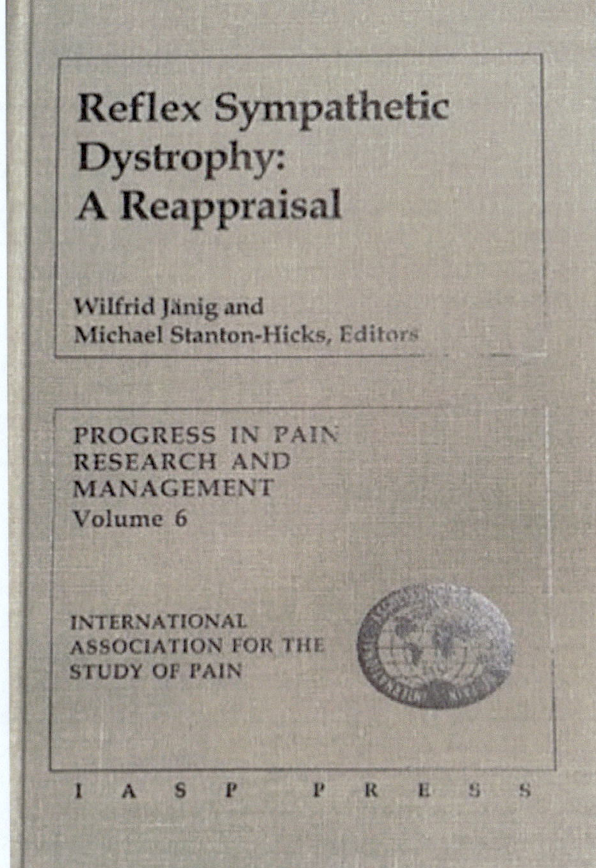

Fig. 2.2 The publication of this book, "Reflex Sympathetic Dystrophy: A Reappraisal" Eds. Wilfrid Jänig, Michael Stanton-Hicks. Progress in Pain Research and Management. Vol 6. I.A.S.P. Press, Seattle, 1996. represents most of the topics that were covered in the Orlando November Consensus workshop and are not a Proceedings of the workshop

2.3   Classification of Complex Regional Pain Syndrome (CRPS)

Reproduced from RA Boas in eds Wilfrid Jänig and Michael Stanton-Hicks, Reflex Sympathetic Dystrophy: A reappraisal: Progress in Pain Research and Management, Vol 6, 1996, IASP Press, Seattle, pp. 79.

CRPS *describes a variety of painful conditions that usually follow injury, occur regionally, have a distal predominance of abnormal findings, exceed in both magnitude and duration the expected clinical course of the inciting event, often result in significant impairment of motor function, and show variable progression over-time.*

2.3.1   CRPS 1 (RSD)

1. Follows an initiating noxious event
2. Spontaneous pain or allodynia/hyperalgesia occur beyond the territory of a single peripheral nerve(s), and is disproportionate to the inciting event
3. There is or has been evidence of edema, skin blood flow abnormality, or abnormal sudomotor activity, in the region of the pain since the inciting event
4. This diagnosis is excluded by the existence of conditions that would otherwise account for the degree of pain and dysfunction

2.3.2   CRPS Type 2 (Causalgia)

The syndrome follows nerve injury. It is similar in all other respects to type 1

1. Is a more regionally confined presentation about a joint (e.g., ankle, knee, wrist) or area (face, eye, penis), associated with a noxious event.
2. Spontaneous pain or allodynia/hyperalgesia is usually limited to the area involved but may spread variably distal or proximal to the area, not in the territory of a dermatomal or peripheral nerve distribution
3. Intermittent and variable edema, skin blood flow change, temperature change, abnormal sudomotor activity, and motor dysfunction disproportionate to the inciting event, are present about the area involved

2.4   Sympathetically Maintained (SMP)

Pain that is maintained by sympathetic efferent activity or neurochemical or circulating catecholamine action, as determined by pharmacological or sympathetic nerve interruption is termed Sympathetically Maintained Pain (SMP) and may be a feature of several types of pain disorder. It is not an essential component of any one condition. Conditions without any response to sympathetic block are, by contrast, designated as having Sympathetic Independent Pain states (SIP) (Treede et al. 1992).

Also discussed at the time was what the impact of introducing CRPS would have on certain patient support groups whose foundations has been based on an old terminology. In this regard, the Reflex Sympathetic Dystrophy Association of America since its inception in 1975 as a charity has been a stalwart supporter of RSD patients and has acted as a proxy for recognition of this medical entity, in particular for treatment benefits at the behest of Workers Compensation Agencies and the Insurance Industry.

While SMP is a phenomenon described in humans and is well described in animal models such as the peripheral neuritis trauma (PNT) and spinal nerve transection (SNT) models. Kim and Chung (1991), or also in the chronic constriction injury (CCI) model, where the consequences are more complex (Wakisaka et al. 1991). The resultant SMP is not due to sympathetic vasoconstriction (Drummond et al. 1991). It may be because there has been a loss of postganglionic neurons, a decrease of retrograde axoplasmic transport or nerve growth factor (NGF). Nevertheless, the SMP soon after the CCI lesion does respond to sympathectomy. In fact, pain after the CCI model is mixed SMP & SIP which changes to only SIP later in the course of the disease. Much of this discussion lead by Martin Kolzenberg, Sam Hasenbusch and Bob Boas centered around the failure of sympathectomy and nerve blocks in patients with CRPS 1 and 2. While many theories including the failure of complete sympathectomy can be explained by recognizing that sympathetic fibers from the contralateral sympathetic chain can innervate ipsilateral structures, it may be that responsiveness (upregulation) to norepinephrine in intact nociceptors might also be the culprit. Nishiyama et al. (1994) demonstrated that the α_2 adrenoceptor is expressed on dorsal root ganglion cells.

Peter Drummond

Martin Koltzenburg

Author and Sam Hassenbusch

**Governor-general of New Zealand
Zealand awarding Order of Merit to "Bob" Boas**

It was very clear to all that the animal data of SMP are extremely important for their relevance to CRPS Type I symptoms. While the activation of sympathetic activity may not induce C-nociceptor sensitization its effect at the sympathetic terminal may be critical. Tracey et al. (1995) and Gonzales et al. (1991) have shown that activation of the α-receptors at the sympathetic terminal evokes prostaglandin synthesis. Work by Drummond (1995) would suggest that norepinephrine responsiveness disappears after inflammation subsides. Pain may also be derived from non-nociceptive afferents and some of these afferents may be sympathetically activated (Price et al. 1989; Roberts and Fogelsong 1988; Roberts and Elardo 1985).

Donald Price

The problem of how to distinguish a placebo response from a diagnosis of SMP galvanized David Haddox who together with Ed Covington made a valiant and successful attempt to explain this concept to the group. One way in which a placebo response can be avoided is to do at least two separate block procedures, Bonica (1953) and more recently cognitive factors would be considered more important than any direct association with a specific type of treatment. The credence of *expectation* is supported by models which include a learning component. Siegel (1982) pointed out that Pavlovian Conditioning is the recognition of a known occurrence or non-occurrence to some unconditional stimulus or a change in its magnitude or duration. The patient's expectation of improvement will also influence the magnitude of a placebo response. This is something that does not happen in isolation, i.e., patients inherently want to have a good response—"they want to please their doctor"—and hope their treatment works (Jensen and Karoly 1991). The use of double-blind comparisons or the use of a hidden infusion in which the active drug is unknown to the patient are ways of controlling for placebo responses. Now that our understanding of the nature of SMP has improved so much, the role of the sympathetic nervous system as an integral component of the various pathophysiological pain syndromes and its seeming absence in certain individual patients has always been questioned because so much reliance rested on the outcome of sympatholytic procedures. Many such problems raised during previous discussions can be resolved by well controlled studies in groups rather than in individual patients.

As was neatly summed up by Martin Kolzenburg and Eric Torebjork, at the end of a very long session regarding animal models, there are many mechanisms that can provide more information responsible for persistent pain and hyperalgesia than those which are associated with inflammation and chronic neuralgia (Koltzenburg and Torbjörk 1995). Such changes may be responsible for persistent catecholamine sensitivity which has already been demonstrated by. Both the central and peripheral nervous systems adapt to nerve injury by developing catecholamine sensitivity and heat sensitization in nociceptors which may persist for months or years. The central nervous system reacts to these changes by signaling hyperalgesia to touch, pinprick and cold. Treatment and its successful response may only occur if such changes can be reversed.

Eric Torebjörk

2.5   Psychological Factors

Psychological processes as fundamental to or as contributary factors could constitute a component of RSD. One report by Van Houdenove (1986), suggested that a significant life event such as a personal loss coincident with the onset of RSD because of its emotional impact could influence the clinical nature of the subsequent pain response. Strong emotions, loud noise, bright light and excitement are triggers of sympathetic stimulation all of which increase pain (Doupe et al. 1944; Bonica 1979). Any autonomic arousal associated with catecholamine release—anxiety or stress reaction (Ecker 1989), would suggest a feedback loop that exacerbated the symptoms of RSD. Bruehl and Carlson (1992) reviewed data in 20 publications which while lacking significant scientific rigor did identify some aspects that are consistent with depression, anxiety or stress that could be channeled through α-adrenergic mechanisms. Some individual behavioralists, previously mentioned, have suggested that a *type A* personality may express heightened sympathetic activity and therefore be more susceptible to RSD (Contrada 1989).

The behavioral responses in RSD are similar to those seen in most chronic pain conditions. Pilowsky has described "illness behaviors" as "abnormal illness behavior," an exaggerated "sick role" (Pilowsky 1969; Pilowsky et al. 1984). As a result of their acquired disuse, overprotection and immobilization, demineralization and vasomotor changes, such patients are plagued by their poor outcome, a consequence of disease duration. Everyone agreed that mild cases will often spontaneously recover, and others will regain function after physical therapy whereas for some, inactivity will adversely affect both function and recovery. White and Sweet (1969) stated that trophic sequelae are more likely due to prolonged disuse of the affected extremity caused by pain, as a consequence of which physiological changes and cutaneous dysesthesia will develop in such patients. "Patients who are poorly motivated to get well and are unwilling to make the effort needed to overcome the discomfort and stiffness that follows immobilization after trauma," is an opinion that pays no credence to the fact that such trophic sequelae *may* also incur from CNS dysfunction. Hemler et al. (1988) who evaluated servicemen with previous nonspecific symptoms found no evidence for a prevalence of these characteristics having an influence on the reflex or inflammatory mechanisms of the clinical entity. In fact, he promoted a multimodal interdisciplinary approach that addresses both the reflex and inflammatory mechanisms of the disorder. In children, Lynch (1992) noted that premorbid characteristics involve parental conflict, hyper-responsibility, and non-assertiveness in most cases. Haddox described his use of the Dartmouth pain questionnaire and the State Trait Anxiety Inventory to compare patients with radiculopathic symptoms but found no differences in outcome between the two groups. Furthermore, Deleo et al. (1983) using Eysenck's Personality Inventory also found no differences between patients with Sudeck's atrophy from those with other neuropathic pain symptoms. At the end of an exhaustive debate on psychological aspects, it was acknowledged that the so called "RSD personality" is a *myth* and cannot be substantiated. Conversion and dissimilation may masquerade as RSD or even coexist with it in a form of conscious or unconscious exaggeration. Such cases may tax any well-meaning clinician.

While the place for regional anesthetic blocks including sympatholysis as an aid to diagnosis of RSD has had a long history, their use during World War II in service personnel described by Doupe et al. (1944) and supported by the views of Livingstone (1943) and Leriche (1949), both of whom considered sympatholysis at the time to be necessary for a diagnosis of RSD. However, limitations and the relationship to placebo effects are already well described in this text but as a procedure to facilitate rehabilitation their role in a treatment plan can, to this day be justified. The control and execution of regional anesthetic procedures include adequate training, technical skill, and the need to repeat the procedure whenever the expected outcome fails to materialize. Successful sympathetic block, whether in the cervical region—stellate ganglion or cervical ganglion block; the efficacy of lumbar sympathetic block requires adequate monitoring of its technical response, namely, temperature control in the subtended region. In the absence of severe arteriosclerotic vascular disease or other vascular pathology, a final temperature of 35 °C (≥90% interruption of sympathetic supply to the involved region) would also be indicative of a significant SMP component. A smaller increase in temperature of, for example, 30 °C might suggest an element of SIP, local anesthetic factors, or even technical deficiency. For other regional anesthetic procedures such as a somatosensory and/or motor block is required, careful sensory and muscle testing is important. Covington summarizes the discussion of psychological factors below:

"Ed" Covington

1. There is no good evidence that true RSD is a psychogenic condition
2. It is highly likely that anxiety, stress, and chemical dependence increase nociception in RSD. Appropriate treatment with relaxation and antidepressants should help
3. The severe pain of causalgia is the cause of psychiatric suffering and not the converse
4. Pathologic signs and symptoms in RSD may be worsened by volitional or inadvertent behaviors, such as immobilization and disuse. These behaviors may be motivated by fear, misinformation, or aggressive urges. Thus, although maladaptive they may not indicate psychopathology.
5. Patients with conversion disorder and factitious illness may be erroneously diagnosed with RSD and thus received inappropriate treatment. Their failure to respond may mislead professionals into the belief that RSD is a psychiatric condition.
6. Conversion/factitious mechanisms may be more likely to mimic RSD than other pain syndromes. If true, such cases should be relatively rare in the primary care setting and increasingly common in tertiary care and beyond.
7. We lack definitive information as to whether aberrant behavior following a trivial injury can produce RSD that otherwise would not occur. We also do not know whether high levels of state or trait anxiety can do so.

The case for Differential Nerve Block, whether subarachnoid or epidural has been studied for years. Both Prithvi Raj and Gabor Racz strongly supported the continued use of differential epidural anesthetic procedures as a means of distinguishing between autonomic/sympathetic contributions to a patient's pain from nociceptive pain and placebo effects.

They both supported the exclusive use of differential epidural vs. spinal block procedures because of safety and practical aspects. The epidural route has greater utility as it can be undertaken at any spinal level.

The local anesthetic sequence is to use saline followed by one of the short acting local anesthetics like 2-chlorprocain, 0.3–0.15%.

The foundation for these diagnostic procedures was determined from the original studies by Gasser and Erlanger (1929) who noted that when cocaine was applied to the dog saphenous nerve the compound action potential (CAP) of small fibers failed before it disappeared in large fibers—a differential rate of block response. Later, Nathan and Spears (1961) found that when local anesthetics at critical local anesthetic concentrations were applied to cats' spinal nerve roots that the small, myelinated fibers were blocked first, leaving the large fibers unblocked, a phenomenon to which they referred as "absolute differential block." Further studies by Franz and Perry (1974) narrowed this effect down to the requirement that it was necessary to block more than 4 mm of nerve if they were to achieve a differential rate of block. In 1980, Gissen et al. (1980) determined that once equilibrium between a nerve and a local anesthetic was reached, a differential rate of block of C (G IV) fibers were always blocked before the A-alpha fibers. This explains why differential nerve blocks, whether subarachnoid or epidural must be interpreted with caution. If the patient's symptoms are relieved after anterograde (using saline before the local anesthetic) administration of the local anesthetic the pain is considered to be psychogenic. It must also be remembered however that 30–35% of patients with true organic pain will often achieve relief from an inactive agent—a placebo response. The argument for retaining differential nerve blocks as an adjunct to a diagnosis of CRPS was strongly defended by a Gabor Racz, Prithvi Raj and Michael Stanton-Hicks, above. The technique, however, must be meticulously performed if the results are to be interpreted correctly.

The foregoing account completes discussions and recommendations from the Orlando meeting and sets the stage for a meeting of basic scientists and clinicians to develop an algorithm that would be clinically useful for a wide range of practitioners. While a time frame was difficult to envisage, it was hoped that this could take place within the next 3 to 4 years.

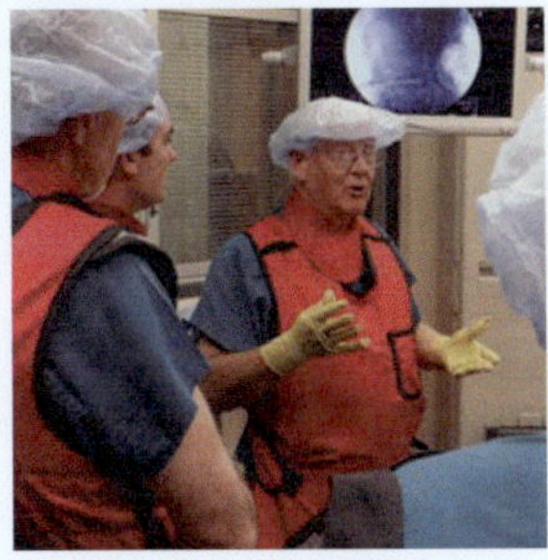

Gabor Racz

The 2-day workshop of 29 faculty was held at a small facility, the Malibu Bay Inn in conjunction with the American Pain Society's (APS) Annual Conference in Los Angeles under the auspices of the SIG, Pain and the Sympathetic Nervous System, IASP during the summer of 1997 to continue the work already begun in Orlando.

> Louisa Jones, Executive Officer, IASP provided invaluable help in solving many of the logistical issues related to IASP and the workshop organization. Most faculty from the last meeting in Orlando were joined by additional members whose contribution was considered invaluable to the outcome of the Malibu project. Some of these faculty were also attending the APS meeting.
>
> The Dahlem-type protocol used at Schloss Rettershof, Stanton-Hicks et al. (1995), was maintained at this workshop. To many of the "original gang" who were at Orlando, Malibu felt like that meeting had never ended and we were about to return to the conference room to pursue our analyses of both the diagnostic criteria and treatment algorithms that were—*works in progress*—after the break! The formal meeting was preceded by a discussion of a hypothetical clinical case study in which 40% of patients with diabetic neuropathy could meet the IASP criteria for CRPS, i.e., 40% had mechanical allodynia, 39% temperature asymmetry and 28% edema, all of which might have been misdiagnosed if the diabetic pathophysiology had been overlooked. The rhetorical question: "is the mere presence of signs and/or symptoms sufficient to satisfy these diagnostic criteria or are the CRPS criteria inadequate, in which case what important clinical signs/symptoms have been omitted?" (Stanton-Hicks et al. 1995, 1998). As suggested by Merikangas and Frances (1993), the ultimate advantage of the CRPS criteria cannot be appreciated until such questions are answered, although a later piece of the formal program, namely validation of the diagnostic criteria was brought up for discussion by Norman Harden. Similar statistical pattern recognition methods like *factor analysis* and *cluster analysis* that were used to validate headache diagnostic criteria (Diehr et al. 1982; Drummond and Lancew 1984; Bruehl et al. 1999) and psychiatric diagnostic criteria (Maes et al. 1992) were the basis for our projected validation studies (Harden et al. 1999) (Fig. 3.1; Table 3.1).

Apart from the two basic scientists and a neurophysiologist, all of the group are clinicians representing medical disciplines that are associated in some way with the management of patients suffering from CRPS. The aim of this workshop is a continuum of the two earlier workshops in addition to moving the frontiers of diagnostic criteria, taxonomy and their place in medical practice and any clinical questions concerning the

Fig. 3.1 Attendees at the Malibu Workshop. Prithvi Raj insisted that suits, ties, or jackets be worn for the "photo shoot" with only three outliers abstaining! Front row, l to r: Prithvi Raj, Donald Price, Wein-hsein Wu, Michael Stanton-Hicks, Bob Boas Norman, Robert Wilder, Harold Merskey, Nelson Hendler Second row: Peter Wilson, Gabor Racz, Torsten Gordh, Angela Mailis, Susan Ford, Medtronic, Gunnar Olsson, Martin Koltzenburg, Nevil Sethna, Christopher Glynn, Giancarlo Barolat Back row: Richard Rauck Wilfrid Jänig, Ralf Baron, Sam Hassenbusch, David Niv, Joshua Prager, John Oakley, Ed, Covington

development of a simple algorithmic tool that could be used by a wide spectrum of medical practitioners. Such an algorithm would assist in an orderly approach to the management of RSD patients and one that is based on our current understanding of its pathophysiology and what clinical approaches have proved to be the most helpful in the absence of any mechanistic-based treatment. The Workshop was also designed to address those animal model behaviors that can translate to the signs and symptoms that are seen in clinical practice already described earlier in this account. The Proceedings of this workshop will determine the factors that will be used to validate standardized diagnostic criteria and a pathophysiology that will determine ultimately form the basis for CRPS treatment. The new complex regional pain syndrome, IASP/CRPS criteria developed during the Orlando conference (Stanton-Hicks et al. 1995) will encompass all those pain conditions that are associated with vasomotor and sudomotor disturbances and will replace all of the previous variety of diagnostic schemes (Blumberg 1991; Gibbons and Wilson 1992; Wilson et al. 1996). The lack of empirical validation of the IASP/CRPS criteria has already been questioned by Norman Hardin above—edema, sudomotor signs and symptoms, vasomotor aspects and motor/trophic signs are already included within criterion three of the proposed changes to the IASP/CRPS criteria.

Another missing piece of the puzzle is the relationship of the diagnostic criteria for setting standards of treatment and management of CRPS (Stanton-Hicks et al. 1995, 1998; Merikangas and Frances 1993).

Table 3.1 Participants and their affiliations at the Malibu Workshop, Summer 1997

Giancarlo Barolat, MD, Division of functional Neurosurgery, Thomas Jefferson University, Philadelphia, Pennsylvania, USA

Ralf Baron Dr med, PhD Klinik für Neurologie, Christian-Albrechts Universität, Kiel, Germany

Robert Boas, MB; BS Pain Clinic, Auckland Hospital, Auckland, New Zealand

Edmond Covington, MD chronic Pain Rehabilitation Program, the Cleveland Clinic Foundation, Cleveland, Ohio, USA

Christopher Glynn, MB; BS Oxford Regional Pain Relief Unit, Oxford, England

Norman Harden, MD Department of Physical Medicine and Rehabilitation, Northwestern University Medical School Center for Pain Studies, Rehabilitation Institute of Chicago, Chicago, Illinois, USA

Samuel Hassenbusch, MD, PhD Anderson Cancer Center/neurosurgery, University of Texas, Houston, TX, USA

Wilfrid Jänig, Dr. med, PhD Physiologisches Institut, Christian-Albrechts Universität, Kiel, Germany

Martin Koltzenberg, Dr. med department of neurology, University of Würzburg, Würzburg, Germany

Phillip Low, MD Autonomic Disorders Research Center, Department of neurology, Mayo Clinic, Rochester, MN, USA

Angela Mailis, MD, MSc Comprehensive Pain Program, Toronto Western Hospital, Toronto Western Research Institute and Department of Medicine, University of Toronto, Toronto, Ontario, Canada

Harold Merskey, DM Department of Psychiatry, the University of Western Ontario, and Department of Research, London Psychiatric Hospital, Ontario, Canada

David Niv, MD Multidisciplinary Pain Control Unit and Pain Research Laboratory, Sourasky Medical Center, Tel Aviv, Israel

John Oakley, MD Yellowstone Neurosurgical Associates, Billings, Montana, USA

Gunner Olsson, MD, PhD Department of Pediatrics and Pain Service Karolinska University, Stockholm, Sweden

Donald Price, PhD Anesthesiology Department Medical College of Virginia, Richmond, Virginia, USA

Gabor Racz, MD department of anesthesiology, Texas Tech University Health Science Center, Lubbock, TX, USA

Prithvi Raj, MD Pain Medicine Center, Los Angeles, California, USA

Richard Rauck, MD department of anesthesiology, Bowman Gray School of Medicine, Winston-Salem, North Carolina, USA

Michael Stanton-Hicks Department of Pain Management, Cleveland Clinic, Cleveland, USA

Robert Wilder, MD, PhD Department of Anesthesia, Children Hospital, and Department of Anesthesia, Harvard Medical School, Boston, Massachusetts, USA

Peter Wilson, MB; BS, PhD Pain Clinic, Mayo Clinic, Rochester, Minnesota, USA

Wen-Hsien Wu, MD pain management center, University of Medicine and Dentistry New Jersey, New Jersey Medical School, Newark, New Jersey, USA

CRPS/RSD in children already acknowledged in a few publications and discussed in Orlando became a centerpiece of the session on pediatrics with Robert Wilder, Gunnar Olsson and Ed Covington leading the conversation. Although prior to the 80's, there had been few reports of CRPS/RSD in children. At this gathering several papers were presented by Bob Wilder and Gunnar Olsson (Olsson et al. 1986; Kesler et al. 1988; Sherry and Weisman 1988; Wilder et al. 1992; Stanton et al. 1993); all of whom have described the symptom complex in children and adolescents. In contrast to the expression of CRPS in adults, is the fact that the lower extremity is affected more frequently than the upper by 5:1, and there is a preponderance of female cases (Sandroni et al. 1998). A mean age of 12.5 years with a range of 3 to 18 is reported. While many of the clinical features are similar to those seen in the adult, children almost universally have spontaneous pain, mechanical allodynia, signs of autonomic dysfunction, particularly edema and a decreased temperature. Color changes may be apparent in about half of children (Sandroni et al. 2003). Because of their exaggerated behavior, children are often considered to have some psychological disturbance or conversion reaction. Behavioral stressors in relation to parents, home or school problems are common. Even if the management of these RSD cases is undertaken in conjunction with a comprehensive rehabilitation program, the interplay of family dynamics and their trigger for the behavioral responses manifested in children embody the axiom, "treating the entire family" could not be more germane. Arner (1991) suggested the use of phentolamine not only as an aid to diagnosis but also to predict the effect of a sympathetic block—a 100% positive predictive value. The impact of RSD on schooling is well documented in the United States with the average number of days lost in a year being at least 40 (180-day school year). Good physical treatment and psychological support, preferably in a comprehensive rehabilitation program can reduce this to 5 days or less per year. Treatment at the end of a structured rehabilitation program should not only extend to the home environment but should also focus on the reintegration of children into school. This is not to say that there will be a few cases in which the foregoing management will fail. Such children with chronic CRPS have similar features to adults in which dystrophic signs and symptoms together with loss of functional use of the affected extremity will still occur in a very small minority despite all treatment efforts. The term, "cold (chronic) CRPS" is axiomatic for adult and child alike.

Gunnar Olsson

At first sight, children may respond to simple physiotherapeutic measures and depending on the severity and impact of CRPS on their physical and mental faculties, it may be necessary to prescribe some form of pharmaceutical support for their initial treatment. In more acute or florid cases, the excellent response obtained after use of an oral steroid (often, a single dose) may in fact induce a complete resolution of CRPS (Christensen et al. 1982; Braus et al. 1994). At times it may be necessary to use tricyclic antidepressants, but their successful response is severely limited by their cholinergic side effects. Desipramine or Trazodone are two of the better tolerated agents. SMP as a factor gave way to much debate (Dworkin et al. 2003). Most of the group did not advocate the use of a sympathetic blocks or other interventions as the first line of treatment. However, almost everyone felt that such measures might be helpful if conservative measures failed (Wilder et al. 1992; Kimura et al. 1995; Lee et al. 2002). Several of the interventionists in the group including this author believe that a sympathetic block during the acute stage can be useful to determine how much SMP might play out on the expression of CRPS. A positive response would suggest a trial of an α1-adrenoceptor blocking agent like terazosin or prazosin, both of which have a long history of clinical use and are well tolerated. A case for continuous sympathetic block with a catheter or even surgical sympatholysis for intractable cases was also suggested. The unanimous opinion was that any one of these decisions can only be entertained in an interdisciplinary environment that is built around those measures that are important to the restoration of function, both physical and mental (Flor et al. 1992; Guzman et al. 2001; Turk and Melzack 2001; Revicki and Celia 1997).

As the Modis Operandi for Malibu, (1) pursuit of a terminology, (2) refining the diagnostic criteria and (3) completion of a clinically useful diagnostic algorithm were the crux of this meeting. However, development of an algorithm, already begun in Orlando was considered prescient in order to make this new information available to practicing clinicians. The original treatment guidelines that were published in 1998 (Stanton-Hicks et al. 1998), reflected the subsequent efforts of a consensus meeting that was held in Minneapolis 2001 (Stanton-Hicks et al. 2002). This meeting was centered around the three domains, rehabilitation, pain management and psychological treatment. An updated version of the algorithm emphasized a time contingent application of those modalities felt to be most appropriate at that phase of a patient's progress in the rehabilitation pathway and one that is not so restrictive chronologically as were the time constraints of the original algorithm that was discussed in Orlando.

While rehabilitation is the mainstay of patient management, an impediment such as pain or some negative psychological aspect will require reassessment of the patient's motivation and to explore what other factors might be interfering with the patient's improvement. The clinician administering the algorithm must remain flexible having continuous feedback from each patient's progress by monitoring simple markers like range of motion (ROM) which should at first be gentle before the load is increased; a measure of strength and the frequency of repetitive movements are crucial to the return of function (Oerlemans et al. 1999a, b). To avoid engendering too much patient resistance to their treatment or exacerbating the intrinsic myofascial syndrome (MFS) with which CRPS is invariably associated, the therapist may be aided by selective trigger-point injections (TPI), motor point stimulation or TENS therapy as suggested in the algorithm.

The algorithm, see Fig. 3.2 was unique in providing a visual assessment of a patient's progress within a multidisciplinary setting. This topic of the workshop was

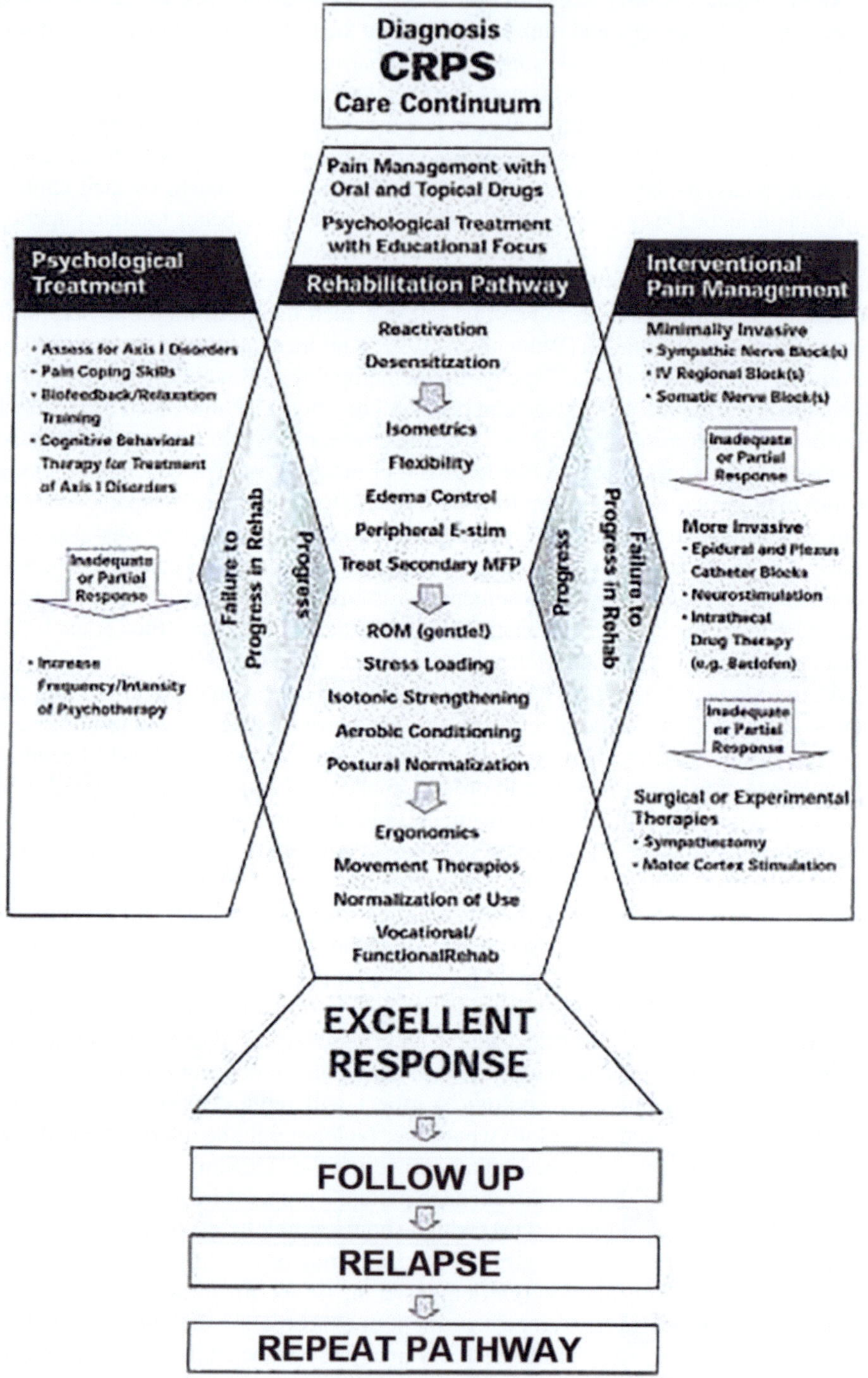

Fig. 3.2 The revised therapeutic algorithm for CRPS. (Adopted from Stanton-Hicks et al. Complex regional pain syndromes: guidelines for therapy. *Clin J Pain*. 1998;14:155–66). Reproduced with permission

enthusiastically endorsed by the whole group. The graded nature of the algorithm emphasized a successive application of specific modalities, psychological, pharmacologic or some intervention that was felt to be appropriate because of a loss of compliance or to address a change in the progress (rate of change) of rehabilitation. The diagnostic criteria already acknowledge that immobilisation is a potential cause or might even precipitate the syndrome whereas a graded approach to reactivation and functional restoration will employ isometric modalities, isotonic strengthening, general aerobic conditioning, partial normalization and "stress loading" (a technique introduced by Lois Carlson, OT to facilitate functional return after hand surgery in patients who had developed RSD), i.e., to mold the impacted extremity into its former self (Watson and Carlson 1987; Butler 2002; Flor et al. 1992). Another aspect found useful for helping a patient to improve in their overall rehabilitation was the introduction of recreational therapy and vocational rehabilitation, an activity that has been particularly well-received by children.

Psychological treatment is based on quality of life (QOL) and helping the patient's coping skills. Cognitive behavioral treatment (CBT) has for years been recognized as an important adjunct to rehabilitation. Where possible, patients should be given an explanation of CRPS, a condition which otherwise from all the social and medical input they experience has only negative connotations. Such candidness will have a salutary influence on their treatment progress if they can be convinced that pain by itself does not mean there is inevitable ongoing damage: rather their adherence to a goal of reactivation of the affected extremity will be an investment in their successful treatment outcome.

Timothy Lubenow

Also relevant to the algorithm is the fact that Axis-1 disorders such as somatoform, mood disorders and schizophrenia may require special psychologic measures and/or treatment with tricyclic antidepressants with or without cognitive behavioral therapy (CBT). Pharmacological support is an integral feature of the algorithm (Christensen et al. 1982; Braus et al. 1994; Kingery 1997). Considerable enthusiasm was expressed by all that the instrument they were responsible for creating would be useful for the practicing clinician. Regional anesthetic procedures and neuromodulation using either spinal cord stimulation or peripheral nerve stimulation are supported by an expanding bibliography and an increasing number of prospective, randomized controlled trials (Hord et al. 1992; Kemler et al. 2000; Oakley and Weiner 1999). Neuromodulation in its various forms is analogous to symptom-specific medications (e.g., muscle relaxants for MFPS) and are crucial for the effectiveness of physicotherapeutic maneuvers by addressing the pathophysiology to the best of our current understanding and to treat the different types of pain such as SMP (Oerlemans et al. 1999a, b; Rauck et al. 1993; Geertzen et al. 1994; Zollinger et al. 1996; Baron et al. 2001). Intrathecal drug therapy also included in the algorithm addresses both analgesia and motor symptoms. The use of intrathecal baclofen for patients who have refractory dystonia is a good example. The best response occurs with dystonia of the upper extremity, the agent being less effective in lower extremity CRPS (van Hilten et al. 2000; Kanoff 1994). In the light of a few positive reports in the literature, surgical sympathectomy was included in the algorithm (Schwartzman and McLellan 1987; Kim et al. 2002; Baron and Maier 1996). To summarize this session, the revised algorithm with the added guide to stimulation techniques was considered relevant for immediate clinical practice. While respecting the final views of the group, Norman Hardin argued that a time contingency for physical methods should be about 12–16 weeks and not the much more restrictive timeframe that was allotted in the earlier algorithm. An interdisciplinary approach with rehabilitation consisting of physical and occupational therapies, pain management and psychological treatment, measured in terms of Quality of Life (QOL) is the optimal approach to patient management. The balance of the meeting was spent tidying up details concerning initiation of the multisite prospective CRPS database that will form the basis for Internal Validation studies of the current IASP/CRPS criteria.

"Norm" Harden "on drums"

Richard Rauck

Cardiff IASP Research Symposium 2000

4

A constellation of researchers drawn from 28 countries organized by the IASP SIG Pain and the Sympathetic Nervous System as a satellite to the IASP World Congress in Glasgow gathered for 1½ days to present their work on basic and clinical science of CRPS to an audience from 32 countries many of whom were attending the main meeting. Bradley Galer as the original instigator of this meeting should be acknowledged.

Bradley Galer

M. Stanton-Hicks, *The Evolution of Complex Regional Pain Syndrome*,
https://doi.org/10.1007/978-3-031-54900-7_4

Heinz-Joachim Häbler

As seen in Table 4.1 each presenter and topic contributed to the 1½ days of meeting discussions. The level of dialogue and progress that was achieved during this symposium was nothing short of spectacular. The international research community has gained a much better understanding of what actually takes place at disparate research sites around the world. To set the stage, Wilfrid Jänig reminded the group that CRPS I is a complex neurological disease that involves the brain and nervous system at many levels. Furthermore, the clinical somatosensory, sympathetic and somatomotor changes that are seen in CRPS I can only be explained by a disease entity that originates in the brain. A number of experimental observations of the central changes that occur in CRPS1 patients by Wasner et al. (2001) already mentioned earlier in this volume have demonstrated that the thermoregulatory reflexes are abnormal in the CRPS I extremity after whole body heating or cooling. The reduction or abolition of cutaneous vasoconstriction in the fingers normally apparent during deep inspiration, already documented certainly suggests the likelihood that an adrenergic mechanism will exacerbate stimulus-evoked pain in CRPS. Drummond et al. (1991, 1994) have suggested that if neurotransmitters are depleted in an allodynic-affected extremity they could evoke supersensitivity of the blood vessels and the expression of α-adrenergic receptors on nociceptive afferents (Arnold et al. 1993; Perl 1999).

Jeanne-Jacques Vatine

Table 4.1 Main topics covered during the first day of the Cardiff research conference

Overview Wilfrid Jänig—CRPS I and CRPS II: a Strategic View

Animal Models—moderator: **Bradley Galer**

Kyungsoon Chung; Sympathetic Involvement in the Spinal Nerve Ligation Model of Neuropathic Pain

Heinz-Joachim Häbler; Neuropathy after Spinal Nerve Injury in Rats: A Model for Sympathetically Maintained Pain?

Wilfrid Jänig; The Value of Animal Models in Research on CRPS

Matthias Ringcamp; the Role of Injured and Uninjured Afferents in Neuropathic Pain

Jean-Jaques Vatine; A Model of CRPS-1 Produced by Tetanic Electrical Stimulation of an Intact Spinal Nerve in the Rat: Genetic and Dietary Facts

Human experimentation—moderator: **Stephen Bruehl**

Ole Anderson; Human Models of Hyperalgesia Induced by Capsaicin: a Discussion of Secondary Hyperalgesia to Heat

Vania Apkarian; Imaging Brain Pathophysiology of Chronic CRPS Pain

Ralf Baron; Human Experimentation

Stephen Butler; Disuse and CRPS

Bradley Galer; Motor Abnormalities in CRPS: A Neglected but Key Component

Angela Mailis; Genetic Considerations in CRPS

Oliver Rommel; Clinical Evidence of Central Sensory Disturbances in CRPS

Jörn Schattschneider; Kinematic Analysis of the Upper Extremity in CRPS

Steven Stanos; a Prospective Clinical Model for Investigating the Development of CRPS

Lijckle van der Laan; The Role of an Exaggerated Regional Inflammatory Response in the Pathophysiology of CRPS

Gunnar Wasner; The Role of the Sympathetic Nervous Autonomic Disturbances and Sympathetically Maintained Pain in CRPS

Roland Wenzelburger; Grip Force Coordination in CRPS

Diagnosis—Michael Stanton-Hicks

Stephen Bruehl: Do Psychological Factors Play a Role in the Onset and Maintenance of CRPS1

Stephen Bruehl; An Empirical Approach to Modifying IASB Criteria for CR PS

Pieter Dijkstra; Reliability of Assessment of Motor Function and Swelling in Patients with Chronic CRPS I of the Upper Extremity

Norman Harden; Diagnosis of CRPS: Summary

Michael Stanton-Hicks; Regional Anesthesia as a Diagnostic Tool for CRPS

Michael Stanton-Hicks; CRPS: Impact of the Change in Taxonomy

Epilogue—Gary Bennett

Philip Finch

Gunner Wasner

To summarise the foregoing descriptions, autonomic dysregulation that we see in CRPS is such an important diagnostic criterion that a single temperature measurement by thermometry taken from the ipsilateral extremity is a sign to record during the differential diagnostic assessment of the patient. Vasoconstrictor activity in the ipsilateral extremity is however only one reason for changes in skin temperature. Such fluctuations are dynamic, changing within minutes or hours. Skin temperature measurement is simple and does not require elaborate measures like thermographic imaging as a monitor, but rather touch, surface measurement or infrared thermometry are perfectly adequate means to document its value. Side to side comparisons are important. In the early stages of the disease, the temperature tends to be warmer than normal but with chronicity it will drop. Similarly, sweating abnormalities that include hyperhidrosis or hypohidrosis are important both from a historical as well as clinical observation. While autonomic dysregulation is just one piece of the pathology that may be necessary for a differential diagnosis of CRPS, so is the symptom of SMP that requires sympatholysis for the measurement of sympathetic function.

Finally, the relationship of autonomic dysfunction and the development of supersensitivity may have a direct effect on the inflammatory process. For example, a deficient blood supply will aggravate metabolic activity, the normal production of substances like prostaglandins and nerve growth factor (NGF) (Tuttle et al. 1993), the results of which such as an increase of nociceptor afferent activity are far reaching. Mast cell stimulation by the increase of NGF will stimulate sympathetic neurons and macrophages to interact with tyrosine kinase (TrkA) receptors which in turn directly sensitize nociceptive afferents (Shu and Mendell 1999). A direct consequence of prostaglandin synthesis is the stimulatory effect on primary afferent nociceptor behavior. Also, contributing to this picture is the activation of α_2 adrenoceptors on sympathetic efferents which also stimulates the synthesis of prostaglandin E2 and I2 (Gonzales et al. 1991). The contribution to pain is therefore multifactorial with the SNS acting both directly and indirectly through the inflammatory process. There is of course little wonder why the SNS has been labelled in the past as the primary culprit in the development of CRPS. The primary focus of pathophysiology has now shifted because of the major advances in our understanding continue to unravel due to the contributions of animal work in contemporary research. Integration of the inflammatory expression and autonomic dysregulation in CRPS provided significant debate amongst the attendees with a plea for more collaborative studies between groups pursuing the two different sciences.

The impact of animal studies to our understanding of motor effects became very clear during the presentation by Jörn Schattschneider and colleagues of their kinemetric analysis in human volunteers (Schattschneider et al. 2001) (Fig. 4.1). These basic motor responses are driven from the premotor area and supplementary motor cortices, having received information concerning the target from the occipital and parietal cortices, they were able to show that constant sensory inputs can change the spatial representation of the affected body region in the

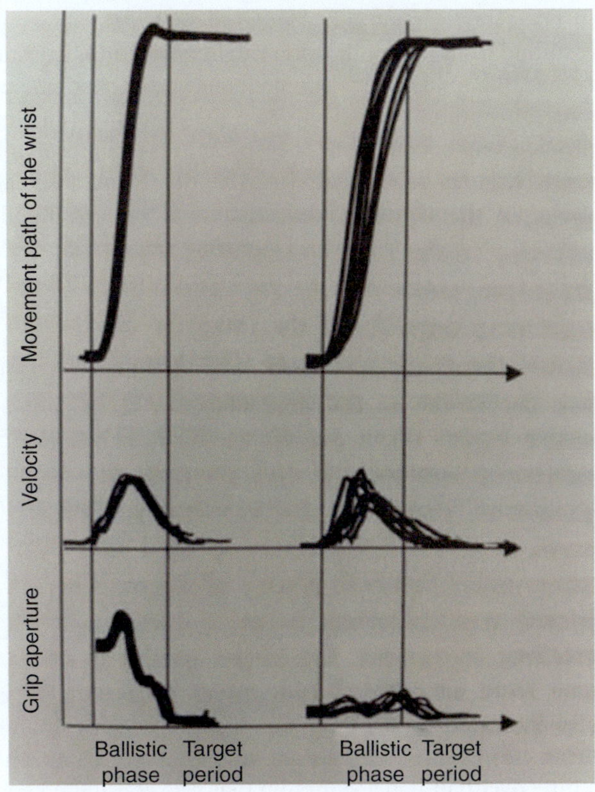

Fig. 4.1 Kinematic analysis from a CRPS patient in comparison with a single control subject. Data show less constant wrist movement paths, peak velocity is decreased and target prolongation in the CRPS patient. Re-printed from Schattschneider et al. Kinematic analysis of the upper extremity in CRPS. In Complex Regional Pain Syndrome, Progress in Pain Research and Management Vol. 22 edited by R. Norman Harden, Ralf Baron, and Wilfrid Jänig. IASP Press, Seattle, 2001 pp. 119–28

primary sensory cortex and subcortical areas (Jaennerod et al. 1995; Merzenich et al. 1983).

Jörn Schattschneider

During this session, Rommel and Thiminer described the general sensory Impairment that occurs in CRPS that this might be a clinical matter of brain plasticity as far as the development of a possible mechanism to explain why such motor defects occur in CRPS? A change in central representation induced by an increase in nociceptive input and decreased sensory, cutaneous, and proprioceptive input resulting from immobilisation might answer this question. As far as therapy is concerned, the reduction of spontaneous and evoked nociceptive input as soon as possible after trauma should reduce these central functional changes (plasticity) ie., physical and occupational therapy are essential prophylactic measures. The patient is advised to engage in activities that do not increase pain while at the same time by using physical therapy and occupational therapy, mobility Is improved, and pain is reduced (Oerlemans 1999). Data from grip force experiments. Associated with the dystonic-like symptoms in CRPS could explain the frequent clumsiness seen in CRPS patients. Had already suggested that sensitization of motor spindles could influence proprioceptive reflexes. Much of our understanding of the motor abnormalities in CRPS remains wanting. Cortical sensory and motor changes after immobilisation of the affected extremity are seen: pain alters motor behaviour and increases motor neglect - postulated by Galer et al, 1995, would suggest a failure of spatial planning and target recognition, although imaging evidence for this assumption is lacking. As an epilogue to the discussion on motor disparities in CRPS, Stephen Butler discussed their work on immobilisation in volunteers. This work was undertaken in the Pain Centre at the University of Washington (U of W) in conjunction with researchers at the University of Uppsala and the Academic Hospital at Uppsala in Sweden who were studying neuropathic pain. Their access to PET scanning gave the U of W researchers the opportunity for comparative observations between the CNS and the periphery. Essentially the study in 20 volunteers nine of whom underwent a PET scan had a fibreglass splint placed on an arm. 78.3% of these subjects had at least one symptom comparible with what is seen in CRPS and a little over half had a symptom that was considered compatible with a neglect-like state. QST studies showed that there was an increase in warm detection to 60.9% and the decrease to cold pain in 52.2%, a finding that also agrees with animal studies that have demonstrated a sensitivity to nonnoxious stimuli because of heating and cooling. Such threshold changes are seen in CRPS 1 patients. The PET studies also identified similar motor phenomena which Schattschneider and colleagues above reported in their kinematic studies. Just how important the comparatively few animal studies of immobilisation in which both sensitivity to sensory stimuli and changes at a spinal cord level have demonstrated is a corollary to the human studies and findings that PET changes observed after patients who were recovering from a stroke could support a "neglect-like" state in CRPS 1 due to immobility alone. A consideration suggested by Stephen Butler is that our current difficulty stems from the fact that that pain may be the root cause of disability and not the reverse. Several prospective human studies were presented during this session. A large prospective (n = 145) cohort of CRPS patients who developed their status after surgery, fracture or soft tissue injuries (Birklein et al. 2000), found that the nature of their pain was deep-seated and of a burning, stinging or tearing nature using descriptors taken from the McGill Pain Questionnaire (MPQ). The sensory disturbances were as follows, 88% hypoesthesia, 53% hypoalgesia, 45% hyperalgesia, 37%. autonomic dysregulation while, 90% of patients demonstrated side to side differences in sweating or skin temperature. Skin temperature was warmer during the acute stage. The group was also reminded that Veldman et al. (1993), had already reported similar signs and symptoms during their large prospective study of 829 post-surgical or trauma derived CRPS patients. Among several studies of total knee arthroplasty (TKR) in which signs and symptoms of CRPS-I that met the criteria in the IASP taxonomy manual (Merskey and Bogduk 1994; Stanton-Hicks et al. 1995) were described by Bruehl et al. (1999) and Harden et al. (1999). Regional pain with dysesthesia, allodynia or hyperalgesia combined with edema and signs of autonomic dysfunction were reported in 69% of patients. MPQ descriptors used included aching 80%, sharp 70% and tender 35%). Of particular note is the fact that 41% of the TKR patients at 3 months and 90% at 6 months met all four CRPS-1 criteria. Because the existing IASP criteria may be more sensitive and less specific, CRPS will tend to be over diagnosed. By separating vasomotor signs and symptoms that include temperature and skin color asymmetry from those changes of sudomotor dysfunction the internal validity could be significantly improved. In a similar manner, by introducing a requirement that at least two of four signs and four symptom categories are included the diagnostic accuracy would be enhanced (Table 5.2).

Ever since Sudeck described algodystrophy a term which became known as Sudeck's Syndrome, the exaggerated inflammatory response after surgery or injury with clinical features of CRPS, languished for some 40 years until research at the University Hospital Nijmegen developed both human and animal models to explore its genesis (Sudeck 1942). Because of the difficulty of obtaining histological evidence, the researchers relied on the examination of amputated limbs in certain cases of CRPS in which pain and contractures had defied all measures of clinical care. Consistent changes of a microangiopathy different from that which is seen in diabetes was noted in the muscle specimens. They suggested that an increase in oxygen-derived free radicals in vascular cells resulted in a disturbed autoregulation of vascular tone. In the chronic phase of CRPS the mixed motor-sensory nerves were largely unaffected. To corroborate their human observations, the researchers developed an animal (rat) model in which a free radical donor tertbutyl-hydroperoxide was used to induce both behavioral and tissue damage responses. The findings included an increased skin temperature, increased paw circumference, redness of the plantar skin and pain behavior all of which support the induction of free radicals that mimic acute signs and symptoms of acute human CRPS. Their hypothesis that CRPS-1 is an exaggerated inflammatory response generated by free radicals, invited its treatment with the antioxidants dimethyl sulfoxide (DMSO) or mannitol infusion that were used successfully in several studies (Goris 1985; Goris et al. 1987; Geertzen et al. 1994; Zuurmond et al. 1996). During the presentation of these data, discussion was somewhat muted as most of those present had been fully engaged with research into the neuropathic nature of CRPS, unfortunately to the exclusion of inflammatory aspects. Since Mitchell's revealing description of causalgia, this aspect of CRPS pathophysiology has dominated discussion and all attempts to tease out other causative aspects of the syndrome. The relatively late observations of Sudeck above were really the first suggestion that CRPS may have an additional component to its pathophysiology that had received little attention at that time. Figure 8.1 is an attempt to graphically visualize how the inflammatory aspect of this syndrome was neglected during the intervening decades. The comparatively recent work by the Nijmegen group, Goris (1985), Veldman et al. (1993), Chaplin et al. (1994), de Mos et al. (2009) and others investigating the inflammatory aspects of CRPS is now integrated with work by other researchers who have been pursuing the neuropathological basis of CRPS. Both aspects of the pathology are equally intertwined in the clinical presentation of this syndrome. As can be seen in appendix X, there has been a dramatic increase in the volume of citations relating to inflammatory, neuropathology and autoimmune aspects of this disease entity. There has been an equally dramatic increase in the volume of investigations into inflammatory aspects of CRPS. Numerous studies of inflammatory and autoimmune aspects are already reported in this text.

4.1 Selected Topics: Generated by the Research Symposium

4.1.1 CNS and Neuropathic Aspects

The central sensory disturbances originally reported by Pette (1927) were reviewed by Mascher (1950) who had reported a case of "Sudeck's Syndrome" with hemisensory impairment in which he concluded that the disease entity can only be caused by CNS dysfunction. These observations have maintained the impression that the etiology is a manifestation of a somatoform disorder, malingering and psychiatric pathology, the contribution to which Verdugo and Ochoa have perpetuated in their description of "pseudoneuropathic pain" (Verdugo and Ochoa 1998). Rommel described his study in which hemisensory impairment measured by the decrease in both temperature and pinprick ipsilateral to the affected limb together with sensory impairment in the upper quadrant in 8/24 patients was measured (Rommel et al. 1999). Mechanical allodynia and motor impairment occurred in a significantly higher percentage of patients which had a generalized sensory impairment but again only in those patients in which sensory deficits were limited to the affected limb. Patients who presented with left-sided CRPS had generalized sensory abnormalities more frequently than those with right sided CRPS. While these findings were interpreted as functional alterations in CNS processing of noxious events that might be a possible source of CRPS pathogenesis, the left handedness could be an anomaly. These researchers postulated that functional disturbances during the processing of nociceptor events occurred in the nucleus ventralis of the thalamus. In another study by the same authors in which sensory impairment in the upper quadrant (3/40 patients) and hemisensory deficits in 12/40 patients was determined, they had more severe pain, greater mechanical allodynia and hyperalgesia and a longer illness than those with spatially restricted sensory deficits. The earlier right-left differences were not substantiated, however.

The value of functional magnetic resonance imaging (fMRI), magnetic resonance spectroscopy (MRS) as additional measures to QST in quantifying objective data were a stimulus for an animated discussion of their ability to elucidate different brain mechanisms that can be associated with chronic pain. Having an opportunity to distinguish differences and similarities within t cortical networks that are involved in various chronic pain conditions gives one a much greater insight when considering the hypothesis that chronic pain is a function of central sensitization mediated by the central nervous system (CNS). This in turn therefore allows us to analyze activation patterns and long-term reorganization occurring in chronic pain conditions. The cognitive processes of pain should, in the future, provide us with targets for future treatment.

Results of magnetic resonance imaging (MRI) of a mixed group of pain patients with headache, back pain, traumatic neuropathic pain and CRPS reported by Thimineur et al. (1998) showed no difference in the CNS pathology other than in those CRPS patients who developed a generalized sensory impairment were found to have more Chiari 1 malformations (8.5%). Recent work using QST also has shown that the sensory impairment occurs in a high percentage of CRPS patients (Rommel et al. 1999, 2001). To summarize this discussion the generalized sensory abnormalities can be correlated with (1) a longer illness, greater pain, (2) a higher incidence of both mechanical allodynia and hyperalgesia and (3) a high incidence of motor dysfunction. Sensory deficits are frequently observed in the early course of chronic CRPS. QST has proved useful to detect sensory deficits together with increased thresholds for cold, warm, heat perception and the sense of vibration ipsilateral to the CRPS limb. Those abnormalities attributed to functional alterations—plasticity seen in both CRPS and other chronic pain conditions were postulated to occur in the nucleus ventralis of the thalamus and the medulla oblongata.

Oliver Rommel

The use of nuclear magnetic resonance spectroscopy (MRS) as an adjunct to fMRI allows for examination of metabolic products in a specific anatomical brain location was elaborated by Apkarian (Apkarian et al. 1995, 1998, 2000). Following such metabolites as γ-aminobutyric acid (GABA), Glutamate (Glu) and *N*-acetyl aspartate (NAA). As opposed to the anatomical images of fMRI, MRS allows one to distinguish long-standing changes of brain chemistry that represent a window into brain circuitry reorganization and is added information to the anatomic changes reflecting instantaneous or acute neuronal activity in the brain. In a study by Stude et al. in which fMRI was used to isolate changes in the cortical representation ipsilateral to chronic CRPS (3–36 mos.) pain 1 h before and 1 h after a sympathetic block provided relief of SMP, a change in NRS of 6.8–3.8, was observed. The reduced blood oxygenation response (BOLD) in the ipsilateral cortical representation maps prior to the block in the affected limb were all increased (Stude et al. 2014). No changes in sensorimotor function were measured. Multiple trials of this nature have shown widespread prefrontal hyperactivity, increased anterior cingulate activity and decreased activity in the contralateral thalamus. The differences identified after sympathetic blocks were performed in CRPS and matched control patients. CRPS patients exhibited a disordered white matter anisotropy and whole-brain grey matter volume. Grey matter atrophy in a single cluster in the R insula, as well as R ventromedial prefrontal cortex (VMPFC) and the R nucleus acumbens and a decrease in anisotropy in the L cingulum-callosal bundle. Such abnormalities underscore the autonomic, emotional and pain perceptions that constitute the wider clinical picture of CRPS (Geha et al. 2008).

Vania Apkarian

To summarize the foregoing remarks, both fMRI and MRS provide us with new insights into chronic pain. The results are complementary to sensory testing, including QST. This information has confirmed that much of the long-term reorganization involving prefrontal hyperactivity in the disease entities of CRPS and chronic back pain can persist for years. These aspects were reinforced by Apkarian as he summarized the benefits of brain imaging. The chemical changes measured by MRS are a long-term consequence of this continuous brain activity and of significance, is a measure of the cognitive impairment that is an invariable component of chronic pain. MRI prefrontal patterns in CRPS and chronic back pain patients suggest that subjective cognitive properties of the pain are unique for each group of patients.

4.1.2   Genetic Aspects of CRPS

Genetic considerations involving the human leukocyte antigen (HLA) class 1 and class 2 molecules of the major histocompatibility complex (MHC) were originally described by Mailis and Wade (1994) in a small group of Caucasian patients. Kemler described similar findings of increased frequency (69%) of HLA-DQ1 in a Dutch cohort of Caucasian patients with IASP criteria for CRPS1, namely pain extending beyond the injury site and Kimura looked at the gene encoding angiotensin-converting enzyme (ACE) (Kimura et al. 2000). This gene is located on chromosome 17 (Kemler et al. 1999). Three genotypes, DD and II homozygotes and ID heterozygotes were identified in the insertion/deletion (I/D) region of the ACE gene. If these genotypes are grouped according to CRPS type-1 and type-2, criteria, 86% of these patients were homozygous for the AC gene deletion and only 22% were homozygous for the ACE gene insertion. The net result is that ACE activity is significantly higher in patients who have a homozygous DD genotype which is therefore a risk factor for CRPS-1. This would suggest that polymorphism testing could be a helpful predictor of neuropathic pain.

Angela Mailis—avocation

Marius Kemler

Further research in animals conducted by Kim and Chung (1992) have revealed a linkage in the mouse homolog of the MHC region, i.e., the H-2 region on chromosome 17, (a mechanical allodynia model). Human Data from recent phantom limb pain studies in which DNA-based-micro satellite typing (short tandem repeats {STR}) markers were used to screen the entire genome have identified an association with a location close to the MHC region—6p21 (Gershon et al. 1999). Such mapping may unravel the contribution between genetics and CRPS. Clearly, both environmental factors either alone or in conjunction with genes can affect complex biological phenomena such as pain (Crabbe et al. 1999). Future research should be engaged in psychosocial, environmental as well as genetic and non-genetic factors that could influence pain perception and treatment outcome.

4.1.3   Impact of CRPS on the Motor System

In an attempt to measure the degree of motor impairment in CRPS Peter Dijkstra and Jan Geertzen used the American Medical Association's Guides to the Evaluation of Permanent Impairment for their study (Geertzen et al. 1998). The term smallest detectable difference (SDD) which is the smallest amount of change in a variable that can be measured with statistical significance ($p \leq 0.05$) was applied to the range of motion (ROM) on the affected and unaffected sides were carried out using shoulder extension/flexion and supination exercises at the elbow and radial deviation at the wrist. 3-point grip, pinch grip and full fist grip strength were measured using a handheld dynamometer and any associated edema was assessed by limb circumference measurements. The limitations of these data in terms of swelling were acknowledged by the group. It was clear that the SDD's for the circumference measurements was extremely variable and exceeded the ipsilateral site in only about 50% patients in comparison with subjective observations (Geertzen et al. 1998). Obviously, the circumferential measurement was not sensitive enough to observe the fine changes of skin texture such as loss of skin creases and other departures from normal skin integrity—which need to be recorded separately. The clinical symptoms of early CRPS vary tremendously and clinical observations of these changes are very important for baseline measurements. Oerlemans et al. (1999a) described a frequent incompatibility between the clinical and laboratory assessments of patients, a reflection of observer unreliability and the subsequent error rate. SDD's have not been used to measure pain of CRPS although van der Kloot et al. (1995) did publish some figures for nonspecific pain syndromes. Physicians should be aware that normal variations within these different types of pain e.g., actual—minimal and maximal—that exceed 1/5 of the total range, that their assessment of ROM, grip strength and edema will invariably be subjective and without scientific accuracy.

4.1.4 "Wrap-up"

As a sequel to the CRPS story, given the early clinical diagnostic experiences of Evans et al. (1946b), Bonica (1954), Kozin et al. (1981), Gibbons et al. (1992), the Schloss Rettershof meeting in 1988 was the genesis for all four consensus conferences and the research symposium, have grappled with the diagnostic criteria and helped to develop a taxonomy that is acceptable to the Committee for Classification of Chronic Pain of the International Association for the Study of Pain (IASP) at the same time showcasing an extraordinary pantheon of contemporary clinical and animal research. All these deliberations have provided a level of standardization and yielded diagnostic criteria that are sensitive and while still lacking a mechanistic basis they do have a specificity that has markedly improved diagnostic accuracy, clinical communication, and direction for research homogeneity (Jänig and Stanton Hicks 1995; Galer et al. 1998a; Bruehl et al. 1999; Harden et al. 1999). Both internal and external validation studies substantially improved their fidelity, i.e., statistically derived diagnostic categories: (1) sensory (hyperesthesia), (2) vasomotor, temperature and/skin color asymmetry, (3) Sudomotor/edema in the affected limb (and/or sweating asymmetry, (4) motor/trophic (reports of motor dysfunction or trophic changes) (Bruehl et al. 1999; Harden et al. 1999; Galer et al. 1998b). The validation studies underscored the fact that when two objective signs from each symptom category are included a very satisfactory sensitivity of (0.70) is yielded while at the same time a specificity of (0.95) Bruehl et al. (1999) is retained. While not yet perfect as a standard test for CRPS, QST can be employed to enhance the clinical sensory testing (Fowler et al. 1987; Hansson 1994; Yarnitsky and Sprecher 1994). Spot temperature measurement preferably with infrared thermometry can provide good and direct evidence of vasomotor disturbance asymmetry (Bruehl et al. 1996). Sweating abnormalities can be tested using QSART 1 (Low et al. 1983). ROM, weakness, dystonia and tremor our assessed. The sympathetic skin response test which may be abnormal in early CRPS generally becomes normal in the chronic phase. Sympatholysis which may suggest SMP can be used to facilitate physical therapy and to test the treatment response to an α_2-adrenergic blocker (prazosin, phenoxybenzamine or terazosin).

In the final analysis, as summed up by Gary Bennett and predicted by Wilfrid Jänig from day one, the advent of animals to the research armamentarium has led to sweeping advances in knowledge. Animal models have made it practical to explore the disease scientifically and while clinical investigations are of necessity limited by what can be done to a human being, animal research in this regard is much less restrictive. There has been a vast and dramatic upsurge in the amount of basic research on the pathophysiology related to CRPS. The pharmaceutical industry has responded with massive investment as the search for potential therapies is explored. These activities into the realm of basic research have opened our eyes to the incredibly complex and varied nature and specific responses that occur in the central and peripheral nervous systems after injury. Our current understanding of the injury response was dependent on a relatively primitive means of measurement as well as a lack of knowledge concerning the involved neuropathology has markedly improved over the past three decades by an order of magnitude. Clinically, it is expected that the pharmaceutical industry will eventually respond with pathology-specific drugs that will improve the treatment of CRPS patients. Because of the multifactorial pathophysiologies inherent in CRPS, there will never be a *one only* pharmaceutical that will effectively treat this disease entity.

The Cardiff Research Symposium has been incredibly revealing, it has buried countless myths and I would suggest that in the future, pathophysiological aspects will ultimately unravel leading 1 day to a fundamental mechanistic characterization of this disease entity.

As a codicil to the story of CRPS, and I believe a testament to the introduction of CRPS which as a term has been a lightning rod for its influence in focusing attention on the core questions, namely, (1) The search for the pathophysiology underlying nerve-injury responses in the peripheral and central nervous systems, (2) A re-evaluation of the role of the SNS in CRPS and (3) rejection of the opinion that CRPS is a psychiatric condition, (4) the large number of clinical studies evaluating different aspects of the clinical presentation, and the even larger number of animal studies trying to tease out the underlying mechanisms (Fig. 4.2).

In his Epilogue to the Research Symposium and the subsequent publication (Progress in Pain Research and Management: Complex Regional Pain Syndrome; Eds. Norman Harden, Ralf Baron, Wilfrid Jänig, 2001 IASP Press, Seattle) of much research material presented at the Symposium, Gary Bennett laid out what the future might look like. To use his phraseology: *"our current store of knowledge is largely an unassimilated collection of facts of unknown clinical relevance. This information must be sorted into some sort of coherent picture. Identifying the mechanisms of clinical relevance will be extremely difficult and, I believe the primary bottleneck to progress."* Looking at a future, the acquisition of information relevant to a possible mechanism involved in CRPS pathophysiology will not be easy. Certainly, help from the pharmaceutical industry will be invaluable and possibly the only way in which some giant steps in this direction can be made. Differential diagnosis is the main conundrum, and one in which refinement of the diagnostic criteria should help improve. At least it can be concluded that changing the name, in this case, Reflex Sympathetic Dystrophy

Fig. 4.2 Most of the scientific content from the Cardiff Symposium is included in the publication, "Complex Regional Pain Syndrome", Eds. Norman Hardin, Ralf Baron, Wilfrid Jänig. Progress in Brain Research and Management, Vol 12, published by IASP Press, 909 NE 43rd St. Suite 306, Seattle WA 98105, USA

Gary Bennett

(RSD) to Complex Regional Pain Syndrome (CRPS) did galvanise a dedicated scientific and clinical community to all head in the same direction. The disparity between the rather narrow way in which this disease entity has been studied over

research pathways

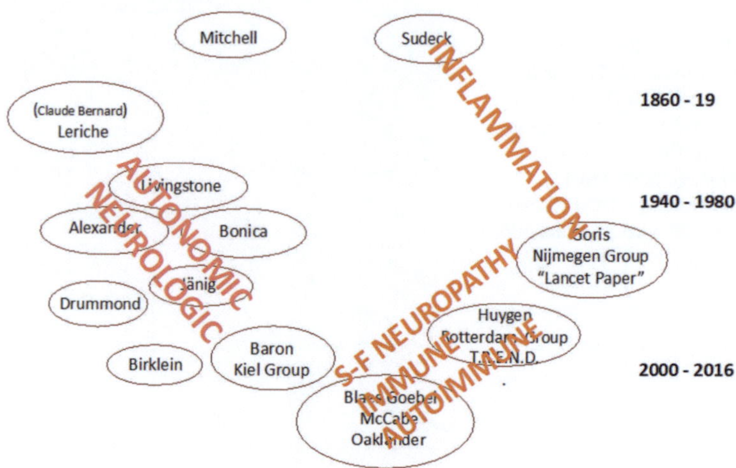

Fig. 4.3 Research pathways from 1860 to the present. M. Stanton-Hicks. The diagram emphasizes two separate pathways—**Autonomic/Neurologic** and **Inflammatory** that incorporated interest and study of a source for the expression of RSD over the period form 1860–2016 and their final combined study activities—**Small-fiber/immune/autoimmune**. The names in oval graphics represent individuals who were either pioneers or leading researchers at site-specific research centers: **Mitchell** categorized Causalgia during the American Civil War. **Sudeck** pioneered the study of radiological inflammatory changes seen in Reflex Sympathetic Dystrophy (RSD) in Germany. **Leriche**, French surgeon influenced by **Claude Bernard**, the French physiologist who was the first to describe pain relief by distal extremity sympatholysis. **Livingstone**, American surgeon who emphasized a pathophysiologic relationship between the sympathetic nervous system and pain in RSD. **Alexander**, contemporary anesthesiologist with **Bonica** emphasized SNS as a source of RSD. **Jänig** pioneer of animal model and translational research in RSD. **Drummond** led the West Australian research team sourcing SNS as a component of RSD. **Birklein**, German neurologist, research into neuropathic pain and other pathophysiologies of RSD. **Baron**, German neurologist who emphasized studies and opinion concerning the neuropathic nature and autonomic aspects of RSD. **Blaes** and others studied immune and autoimmune aspects of RSD. **Huygen**, Rotterdam, pursued many human studies of inflammatory mediators of RSD. **Goris**, and team in Nijmegen, undertook original human and animal studies related to inflammatory aspects of RSD pathophysiology

the decades is seen in Fig. 4.3. This research symposium is evidence of the contemporary widespread interest in seeking out a basis for its pathophysiology.

The figure depicts how the inflammatory aspect of this syndrome was neglected over the intervening decades while neuropathic and vasomotor studies dominated clinical and animal research over the same period.

Budapest, August 2003

5

This was the last and pivotal of all the consensus conferences. The same workshop format used in the three preceding workshops was maintained during this meeting. Now almost four decades on and 15 years since Schloss Rettershof, the proposal that was submitted to the IASP at that time has become the defining instrument.

5.1 Diagnostic Criteria of CRPS: See Comparison of the Criteria from the Two Periods, 1988–2003

Before describing the workshop that was held in Budapest, some additional explanation of the Dahlem structure is indicated. The original Dahlem format described earlier in this text for such workshops involved a discussion of a topic about which each participant knew very little to which each had been assigned other than perhaps a background reading of the literature. The specific difference from a strict Dahlem format which now resides in the Free University of Berlin within the Dahlem precinct, and the format that was used in these workshops is that the *Background Reading* used for breakout discussions was replaced by lecture material that was provided by selected speakers. These dialogues contained the fodder that was used by the members of each breakout group for their deliberations. Individuals and a spokesperson (leader) of dissimilar backgrounds would constitute these groups which were deliberately chosen to reflect differing opinions while at the same time being ignorant of each selected topic. On conclusion of their working sessions, the results of which were then assembled by each leader who then presented the respective summaries to the entire assembly for discussion. On completion of the presentations, a new round of breakout groups given topics already voted upon by the entire assembly were organized after which each respective leader presents their group's consensus views to the entire assembly for debate. The fallout from these endeavors was a unanimous endorsement of the earlier empirical research recommendations (Harden et al. 1999; Bruehl et al. 1999).

© The Author(s), under exclusive license to Springer Nature Switzerland AG 2024
M. Stanton-Hicks, *The Evolution of Complex Regional Pain Syndrome*,
https://doi.org/10.1007/978-3-031-54900-7_5

Fig. 5.1 Participants at the Budapest 2003 Workshop
   First row: Tim Lubenow, Mark Hendrickson, Anne Louise Oaklander, Norm Hardin, Robert Schwartzman, Srinivasa Raja, Gunnar Olsson, Gabor Racz, Harold Merskey
Second row: Richard Rauck, Marshall Bedder,?, Heinz-Joachem Häbler, Marius Kemler, Helmut Blumberg, John Loeser, Michael Stanton-Hicks, Gunnar Olsson, Prithvi Raj, David Niv
Third row: Jörn Schattschneider, Wilfrid Jänig, Joshua Prager, Robert van Hilten, Lijkle van der Laan, Ralf Baron, Frank Birklein, Peter Drummond, Philip Finch

Josh Prager

   The objectives of this meeting during the first day, were to review all of the material that has formed the basis for the diagnostic criteria and its assimilation within the Taxonomy of Chronic Pain. Some 35 faculty from seven countries are shown in the Table 5.1. The Dahlem format allowed the breakout sessions during day 1 to use the lecture material presented as *background reading* which was then brought back

Table 5.1 Participants and their affiliations that attended the Budapest workshop

Ralph Baron Dr Med Klinik für Neurologie, Christian-Albrecht Universität, Kiel, Germany	**Harold Merskey, DM**, Department of Psychiatry, University of Western Ontario, Toronto, Canada
Frank Birklein, Dr med, PhD, Neurologische Universitätsklinik, Mainz, Germany	**David Niv, MD** Tel-Aviv Sourasky Medical Center, Tel-Aviv, Israel
Helmut Blumberg MD Neuroschirurgische Universitäts Klinic, Klinikum der Albert-Ludwigs-Universität, Freiburg, Germany	**Gunnar Olsson, MD, PhD**, Pain Treatment Service-Paediatrics, Karolinska Hospital, Stockholm, Sweden
Stephen Bruehl, PhD, Vanderbilt University, School of Medicine, TN, USA	**Joshua Prager MD**, UCLA Spine Center, Santa Monica, USA
Alan Burton, MD, MD Anderson Cancer Center, TX, USA	**Gabor Racz, MD,** Pain Center, Texas Technical University Health Sciences, Lubbock, USA
Peter Drummond, PhD, Murdoch University, Perth, W. Australia	**Prithvi Raj, MBBS**, Pain Center, Texas Technical Health Sciences
Jan Geertzen, MD, PhD, Center for Rehabilitation, Groningen, Netherlands	**Richard Rauck MD**. Carolinas Pain Institute, Winston-Salem, USA
Heinz-Joachim Häbler, Dr. med, Christian-Albrechts Universität, Kiel, Germany	**Oliver Rommel Dr med**, Rommelklinik, Schmerztherapie und Neurologie, Bad Wildbad, Germany
Norman Harden MD, Rehabilitation Institute of Chicago, USA	**Robert Schwartzman MD**, Department of Neurology, Drexel University College of Medicine, USA
Mark Hendrickson, MD, Department of Orthopedics, Cleveland Clinic, Cleveland, USA	**Michael Stanton-Hicks, MB; BS, Dr med**, Pain Management Center, Cleveland Clinic, Cleveland, USA
Thomas Janicki, MD, Case Western Reserve University, Cleveland, USA	**Lijckle van der Laan, MD**, St Antonius Hospital, Nijmegen, Netherlands
Wilfrid Jänig, Dr med, Christian-Albrechts Universität, Kiel, Germany	**Robert Van Hilten, MD**, Leiden University Medical Center, Leiden, Netherlands
Marius Kemler, MD, PhD, Martini Hospital, Groningen, Netherlands	**Gunnar Wasner, MD**, Klinik für Neurologie, Christian-Albrecht Universität, Kiel Germany
John Loeser MD, Neurological Surgery, University of Washington, Seattle, USA	**Robert Wilder, MD**, Department of Anesthesia, Mayo Clinic, Rochester, USA
Timothy Lubenow, MD, Rush Pain Center, Chicago, USA	**Peter Wilson, MB;BS, PhD**, Spine Center, Mayo Clinic, Rochester, USA

to the entire group for debate following which a second group of breakout sessions provided the final arguments that were discussed at a closing general assembly. During this session, the previous empirical recommendations for research will be replaced by those research criteria that were recommended as a tightened specificity in the diagnostic criteria and therefore, will be a component of the future validation studies that are planned within the next 3 years.

The second day of the meeting was devoted to the treatment algorithm and the integrated way in which selected modalities are incorporated into the functional restoration pathway of CRPS patients. The previously very restrictive time frame for using such modalities was relaxed thereby allowing a particular rehabilitation plan to occupy 12 to 16 weeks which were considered more appropriate. Already described earlier in this text, both Norm Harden and Jan Geertzen had argued that the more restrictive timeframe was not compatible with physiological responses in "sick" muscle (MFPS, dystonic behavior) or behavioral adaptation to rehabilitative measures. This not only allows for a degree of elasticity but also acknowledges that all patients are not equal. The meeting was concluded with a 1/2 day that allowed for the presentation of topics to an open audience, many of which came from the Congress of the World Institute of Pain concurrently held in Budapest (Fig. 5.2).

Although there were no proceedings from this workshop, the entire subject material became the text for the IASP publication "CRPS: Current Diagnosis and

Fig. 5.2 CRPS publication containing most of the material discussed during the Workshop

Therapy" editors, Wilson, Stanton-Hicks, Harden. Progress in Pain Research and Management, Vol 32, 2005, and a subsequent review article, Harden et al. 2007, Proposed new diagnostic criteria for complex regional pain syndrome. *Pain Med.* Doi: https://doi.org/10.1111/j.1526-4637.2006.00169.x

Abstract of above article: *This topical update reports recent progress in the international effort to develop a more accurate and valid diagnostic criteria for complex regional pain syndrome (CRPS). The diagnostic entity of CRPS (published in the International Association for the Study of Pain's Taxonomy monograph in 1994; International Association for the Study of Pain [IASP]) was intended to be descriptive, general, and not imply etiopathology, and had the potential to lead to improved clinical communication and greater generalizability across research samples. Unfortunately, realization of this potential has been limited by the fact that these criteria were based solely on consensus and utilization of the criteria in the literature has been sporadic at best. As a consequence, the full potential benefits of the IASP criteria have not been realized. Consensus-derived criteria that are not subsequently validated may lead to over or underdiagnosis and will reduce the ability to provide timely and optimal treatment. Results of validation studies to date suggest that the IASP/CRPS diagnostic criteria are adequately sensitive; however, both internal and external validation research suggests that utilization of these criteria causes problems of overdiagnosis due to poor specificity. This update summarizes the latest international consensus group's action in Budapest, Hungary to approve and codify empirically validated, statistically derived revisions of the IASP criteria for CRPS.*

During the general discussion, recommendations for and execution of validation studies was provided by Norman Harden. Much of the ensuing discussion wrestled with the rules that should be used to improve specificity while at the same time retaining the degree of sensitivity that will allow most CRPS patients to satisfy the diagnostic criteria, i.e., maintaining a language that was used for the original submission to the IASP Committee on Taxonomy in 1994. These criteria are shown below. As an aside, it is interesting to see how little our original recommendations to the IASP in 1988 differed from the final instrument. Obviously, the inherent standardized, internationally recognized diagnostic criteria for CRPS required statistically driven validation for it to be accepted by the wider research and clinical communities. While still recognising that we have an incomplete understanding of CRPS pathophysiology and therefore do not have a diagnostic "gold standard", the information garnered from each study site will become even more important to satisfy their statistical analysis. The lack of specificity of the 1994 IASP diagnostic criteria which relied so much on symptoms of reported disproportional pain and a historical account of the original injury, interestingly did not include motor signs, ROM and trophic changes.

As a basis for validating our work during the past 2 days these criteria will improve the diagnostic accuracy while at the same time retaining adequate specificity. Two out of four signs and three out of four symptom categories will satisfy the decision rule. Such a change will improve the sensitivity to 0.85 and still retain a specificity of 0.69. The group agreed that such modification while inherently a compromise, should identify as many clinical CRPS subjects as possible while still reducing the false positive diagnostic rate of the current IASP criteria. A summary of the decision rules is shown in Table 5.1. The group agreed that it would be desirable in a research context to identify an improved specificity when such criteria are used for research purposes only. This essentially means there will be two similar sets of diagnostic criteria differing only by their applicable rules.

5.2   Draft of the Definition of Reflex Sympathetic Dystrophy (RSD) for Consideration by the Subcommittee on Taxonomy of the IASP to Replace Supplement 3, 1986 of Pain

5.2.1   Definition

A syndrome of continuous diffuse limb pain, often burning in nature and usually consequent upon an injury or noxious stimulus and disuse, presenting with variable sensory, motor, autonomic and trophic changes; causalgia represents a specific presentation of RSD associated with peripheral nerve injury.

5.2.2   Clinical Features

The symptoms and changes spread independently of both the source and site of the precipitating event, presenting with a glove and sock anatomical distribution. Clinical findings include disturbances of:

Autonomic deregulation: alterations in blood flow, hyper/hypohidrosis, edema
Sensory abnormalities: hypo or hyperesthesia, allodynia to cold and mechanical stimulation
Motor dysfunction: weakness, tremor, joint stiffness
Trophic changes: skin, hair, nails
These features manifest diffusely but not necessarily uniformly in the entire distal extremity. They occur at a variable time after the onset of the syndrome, spreading proximately and occasionally to the opposite side as the syndrome progresses. Dominating symptoms as spontaneous pain, swelling and weakness
Psychological reactive disturbances: anxiety, depression, hopelessness (as with other chronic pain patients)

5.2.3   Diagnostic Tests

Autonomic: Bilateral symmetrical multi digital temperature measures by surface thermistors or thermography, show consistent discrepancies (cooler or warmer) on the affected side.

Sensory: Lowered thresholds to pinprick, light touch and cold.

Motor: Reduced measures of strength as well as active and passive range of motion.

Block response: Sympathetic block usually abolishes diffuse burning pain and allodynia, also raises skin temperature to 35–36 °C see and abolishes the vasoconstrictor response to cold stimulation.

Bone scan: Three-phase scanning shows distinctive, diffuse patterns of increased flow, pooling, and delay.

Nonspecific confirmatory tests: can be used to further quantify changes and follow progress these include radiographic densitometry testing of osteopenia, water displacement plethysmography for measures of swelling, plus tests of sudomotor function, skin conductivity or potentials or quantitative sweat testing.

Staging: cases present with qualitative differences in pain intensity and clinical features, which are not necessarily time or stimulus intensity dependent. Patients with this syndrome should be graded according to the intensity of their presenting features as being mild, moderate or severe in each of the categories of sensory autonomic and motor changes. Eponymous or causative designations provide no quantitative or therapeutic benefit. Patients with sympathetically maintained pain may show benefit to sympathetic block, but have localized pain distribution with only one or two of the other accompanying hallmarks distinctive of RSD.

5.3   Budapest 2003 Draft of the Proposed Clinical Diagnostic Criteria for CRPS Submitted to the IASP Committee of the Classification of Chronic Pain, 2005

5.3.1   General Definition of the Syndrome

5.3.1.1 A Clinical Diagnosis of CRPS Requires the Following Criteria

1. Continuing pain, which is disproportionate to any inciting event
2. Must report at least one symptom in three of the following categories:

    Sensory: reports of hyperesthesia and/allodynia

    Vasomotor: Reports of temperature asymmetry and/or skin color changes and/or Skin color asymmetry

    Sudomotor/edema: Reports of edema and/or sweating changes and/or sweating asymmetry

    Motor/trophic: reports of decreased range of motion and/or motor dysfunction (weakness, tremor, dystonia) and/or trophic changes (hair, nail, skin)

3. Must display at least one sign at time of evaluation in three of the following categories:

    Sensory: Evidence of hyperalgesia (pinprick) and/or allodynia (to light touch and/or temperature sensation) and/or deep somatic pressure and/or joint movement

    Vasomotor: Evidence of temperature asymmetry (>1 °C) and/or skin colour changes and/or asymmetry

Table 5.2 Decision rules

Criterion type	Symptoms for Δ	Signs for Δ	Sensitivity	Specificity
Clinical	3	2	0.85	0.69
Research	4	2	0.70	0.96

For clinical criteria to be realized, two signs and three symptoms must be assessed. For research criteria, to tighten specificity, four symptoms and two signs were agreed by the Consensus Panel. Modified from Table 4. Diagnostic Criteria: The Statistical Derivation of the Four Criterion Factors. In CRPS: Current Diagnosis and Therapy. Progress in Pain Research and Management. Vol 32. Eds. P Wilson, M Stanton-Hicks, RN Hardin. IASP Press, Seattle. 2005. Reproduced with permission

> **Sudomotor/Edema:** Evidence of edema and/or sweating changes and/or sweating asymmetry
>
> **Motor/Trophic:** Evidence of decreased range of motion and/or motor dysfunction (weakness, tremor, dystonia) and/or trophic changes (hair, nail, skin)

4. There is no other diagnosis that better explains the signs and symptoms (Table 5.2)

Because about 15% of all patients previously diagnosed with CRPS might now not meet the diagnostic provisions, using the ICDM format, a CRPS NOS would be added to the terminology to ensure that those patients not meeting the new clinical criteria but whose signs and symptoms could not be better explained by any other diagnosis will therefore receive a clinical diagnosis. The group concluded that once validated by the prospective study mentioned above, the "Budapest" criteria will significantly improve the past problems of medical over-utilization and quality of life (QOL) scores. In addition, they will allow the identification of more homogeneous and pathophysiologically-directed research, the outcomes of which should provide specific direction for future treatment (Janicki 2003; Stanton-Hicks et al. 1998).

Thomas Janicki

The Validation Process. An international, multi-site, between-subjects study designed to make a comparison between the IASP, and Budapest diagnostic criteria should distinguish between the two identified types, CRPS-1 and non-CRPS neuropathic pain patients (non-CRPS pain). 113 CRPS-Type I patients and 47 patients with non-CRPS pain were evaluated at each of the study sites. To maintain study homogeneity any CRPS Type-II patients were excluded. Most of the patients enrolled either had presented with a fracture or had recent surgery. Due to logistical circumstances, the study was delayed by 6 years. The following sites chosen were Reuth Medical Center, Israel, University of Erlangen, Germany, VU University Medical Center, Netherlands, University Medical Center Mainz, Germany, Rush University Medical Center, US, Rehabilitation Institute of Chicago, US, and Leiden University Medical Center, Netherlands. To ensure uniformity of data collection, a previously employed vehicle for multi-site work incorporated a standardized assessment for signs and symptoms (Harden et al. 1999; Bruehl et al. 2002) was used. Each investigator was carefully coached concerning the way data should be extracted.

This meeting sealed the place of CRPS in the contemporary vocabulary of pain physicians and while focusing attention on the clinical aspects of this syndrome it provided the necessary stimulus for those physicians who have an interest in initiating studies which by their nature would either substantiate or refute the premise on which the diagnostic criteria are based. Now that a number of avenues have been opened basic research into the mechanistic aspects of CRPS have taken off in an exponential manner.

5.4   Collated Bibliography and Contemporary Avenues of CRPS Investigation

The question posed on conclusion of the Schloss Rettershof meeting was, "can experimental research contribute to the understanding of reflex sympathetic dystrophy and its related syndromes in order to improve both diagnosis and therapy?" There was a unanimous affirmative response, the intricacies of which required both qualification, experimentation, and time. The way future research will yield information regarding pathophysiology will depend on an understanding of the physiology of the different organs that are clinically impacted by the syndrome. An intuitive emphasis on the development of animal behavioral models that could address the acknowledged characteristics of chronic inflammation in nerve lesions, central lesions, reactive or intrinsic vascular changes, its pharmacological response and of course those specific lesions that are found in a single neuron, motor neurons, sympathetic neurons, dorsal horn neurons and the findings in higher centers representing the entire nervous system are all required.

As a sequel to Budapest, Cardiff and all the previous workshops, our attempts to define these syndromes more accurately is already changing contemporary thought and the approach to managing CRPS. A similar increase in basic science activity, witness the exponential growth of published research studies and the number of clinical reports describing management and treatment protocols related to CRPS.

It was just 5 years since their landmark study categorizing RSD patients (Veldman et al. 1993), that the Nijmegen group then demonstrated peripheral nerve degeneration and muscle abnormalities—chronic inflammation in an autopsy study (van der Laan et al. 1998). Many of their findings were corroborated by Albrecht et al. (2006) who, like van der Laan et al., also noted oxidative stress in amputated limbs of patients with CRPS. Subsequent biochemistry of the tissues revealed the breakdown of phospholipid membranes into malondialdehyde and production of lactic dehydrogenase. While the oxidative stress promoted treatment with antioxidants and free radicle scavengers, the rhetorical question remains, "is oxidative stress a cause or is it a consequence of one or more mechanisms underlying CRPS I?"

Lijckle van der Laan

Norepinephrine hypersensitivity and slow flow/no-reflow influence on inflammation and the metabolism in deep tissues like muscle and bone—and the stimulation and production of inflammatory cytokines. Furthermore, the peripheral nerve lesion that has been demonstrated by Oaklander and colleagues, vasa nervorum with downstream negative effects on axonal nutrition.

Many lines of research have taken advantage of the CPIP model. Recent work in which cerebral proteins were analyzed showed that cerebral involvement occurs in CRPS (Nahm et al. 2014), and from the same group an increased cerebral nuclear factor Kappa B (NF-κB) in CRPS which has a relationship to peripheral nerve injury was detected. An understanding of electroacupuncture treatment has benefited from the use of the CPIP model as the study by Wang et al. (2020) showed that mechanical allodynia is attenuated by RNA sequencing analysis in which multiple genes and gene networks in the DRG are modulated. In another application, the role of endothelin in nociception has been evaluated in CPIP mice. Of interest is the fact that analgesia (anti-allodynic) is achieved by the use of endothelin B-R (ETB-R) agonists. Receptors were found in the plantar muscles but not skin.

By observing changes in the dorsal horn, Chen et al. (2020) demonstrated that the extensive inflammatory and immune responses that occur in the spinal cord dorsal horn (SCDH) during CPIP-induced CRPS1 relate to not only glial activation but also in the upregulation of NLRP3 inflammasome and a large number of differentially expressed genes. By pharmacologically blocking of NLRP 3 inflammasome, IL-1β, glial activation and mechanical allodynia are reduced. The implications for clinical treatment based on this research are obvious.

The intrinsic vascular changes, the pharmacological response and of course those specific lesions that are found in a single neuron, motor neurons, sympathetic neurons, dorsal horn neurons and the findings in higher centers representing the entire nervous system are all required.

As a sequel to Budapest, Cardiff and all the previous workshops, our attempts to define these syndromes more accurately is already changing contemporary thought and the approach to managing CRPS. A similar increase in basic science activity, witness the exponential growth of published research studies and the number of clinical reports describing management and treatment protocols related to CRPS.

It is just 10 years since their landmark study categorizing RSD patients (Veldman et al. 1993), that the Nijmegen group then demonstrated peripheral nerve degeneration and muscle abnormalities—chronic inflammation in an autopsy study (van der Laan et al. 1998).

5.5 Development of Translational Animal Models to Replicate the Clinical Features of CRPS

5.5.1 Animal Chronic Post Ischemia Pain Model (PCIP)

One of the four avenues of animal behavioral research that will be discussed is the chronic post ischemia pain (CPIP) model that was developed by Coderre et al. (2004). By producing ischemia in the rat limb with a torniquet on the ankle, free radicals, oxidizers and enzymes are produced. On release of the tourniquet, reperfusion is associated with the discharge of free radicals—per hydroxyl—and others which in turn produce an ischemic injury to the microvascular cells causing leakage, edema and norepinephrine hypersensitivity and thus a role for activation of the sympathetic nervous system in CRPS. Using the CPIP model Tang et al. (2017) have explored sex differences during inflammatory states in mice. The results have suggested that sex-specific nociceptive hypersensitivity in CRPS I may be dependent on these inflammatory and oxidative stress states. This information could influence translational studies that might benefit the treatment of patients with CRPS1.

Microvascular dysfunction has been studied for decades by Jänig (1988), Blumberg and Jänig (1983), Devor and Jänig (1981). The ischemia/reperfusion (I-R) Injury that is produced by this animal model Increases norepinephrine hypersensitivity and slow flow/no-reflow influence on inflammation and the metabolism in deep tissues like muscle and bone—and the stimulation and production of inflammatory cytokines.

Furthermore, the peripheral nerve lesion that has been demonstrated by Oaklander and colleagues, Oaklander et al. (2006) could also be a consequence of the slow flow/no-reflow phenomenon at the level of the vasa nervorum with downstream negative effects on axonal nutrition.

Many lines of research have taken advantage of the CPIP model. Recent work in which cerebral proteins were analyzed showed that cerebral involvement occurs in CRPS (Nahm et al. 2014), and from the same group an increased cerebral nuclear factor Kappa B in CRPS which has a relationship to peripheral nerve injury was detected. An understanding of electroacupuncture treatment has benefited from the use of the CPIP model as the study by Wang et al. (2020) showed that mechanical allodynia is attenuated by RNA sequencing analysis in which multiple genes and gene networks in the DRG are modulated. In another application, the role of endothelin in nociception has been evaluated in CPIP mice. Of interest is the fact that analgesia (anti-allodynic) is achieved by the use of endothelin B-R (ETB-R) agonists. Receptors were found in the plantar muscles but not skin.

By observing changes in the dorsal horn, Chen et al. (2020) demonstrated that the extensive inflammatory and immune responses that occur in the spinal cord dorsal horn (SCDH) during CPIP-induced CRPS1 relate to not only glial activation but also in the upregulation of NLRP3 inflammasome and a large number of differentially expressed genes. By pharmacologically blocking of NLRP 3 inflammasome, IL-1β, glial activation and mechanical allodynia are reduced. The implications for clinical treatment based on this research are obvious.

5.5.2 Animal Fracture-Cast-Immobilization Model (FCIM)

Because fracture and other orthopedic injuries of the limbs are a common denominator in the genesis of CRPS1, development of an animal model to simulate such events was first reported by Guo et al. (2018).

This fracture-cast-immobilization model (FCIM) replicated many of the signs and symptoms that are seen during the onset of CRPS—abnormal nociception, inflammatory symptoms, dyskinesia, nutritional and cognitive changes. Other animal models have opened up the opportunity to use the FCIM for longer experimental periods. Birklein et al. (2018) found for example that induced mechanical allodynia may last 5 months. After 4 weeks of immobilization in healthy human volunteers the changes in skin temperature, mechanical pain sensitivity and cold pain sensitivity, are reproducible (Terkelsen et al. 2013). Similar changes are seen in rats with limb immobilization. Hindlimb CRPS-like symptoms are associated with increased concurrent skin levels of substance P (SP), calcitonin gene related peptide (CGRP) and the NK1 receptor for SP. In addition, keratinocyte proliferation and inflammatory mediator expression is observed in the hind paw and C-Fos was activated in the spinal cord, all of which support the contention that just like fracture and other trauma, immobilization contributes to the development of CRPS1. In a study using healthy rats that were anesthetized and cast for 2 weeks to create chronic pain post-cast pain (CPCP), all developed local inflammation, tactile and cold allodynia, thermal hyperalgesia, mechanical hyperalgesia for 10 weeks. These findings have encouraged other research groups to use this specific model for chronic translational experiments, particularly those modelling CRPS.

Frank Birklein and son

5.5.3   Animal Passive Transfer Model (PT-TM)

Much recent work by Goebel et al. (2011) and Goebel and Blaes (2013) has supported the contention that CRPS might have autoimmune features. The passive transfer trauma model (PT-TM) which was developed by Tekus et al. (2014) has been used to test this hypothesis. The role of an Interlukin-1-induced mechanism to transfer CRPS by human antibodies into mice was demonstrated by Helyues et al. (2019). This model has opened a huge opportunity to study autoimmune aspects of CRPS in a manner that is similar to the one that has been used to look at the origins of Myasthenia Gravis and pemphigus. Here, the pathological changes after the injection of IgG In humans were replicated in animals. Goebel and Blaes (2013) have suggested that CRPS might be a new autoimmune prototype naming it, an injury-triggered regional-restricted. Autoantibody-mediated autoimmune disorder with a minimally destructive course (IRAM), their premise being that trauma induces pathogenic autoantibodies against nervous system structures. Clinical features of CRPS, edema, mechanical hyperalgesia and SP are significantly increased.

By extension, the TFC-IM model has been extended to include the Passive Transfer Model (PTM). This allows two experimental applications to be used, a mouse-mouse and a human-mouse mode respectively.

Interestingly, it is the pro-nociception immunoglobulin sub type IgM? and not IgG which may have a specific relationship to the pathophysiology of CRPS I. Rodents that received injections of patient IgG all developed enhanced mechanical hyperalgesia in the injured limb. These results would suggest that pathogenic autoantibodies can develop activity early in the context of an injury, a finding that is consistent with post traumatic CRPS. What is not clear however, is whether CRPS-related IgM autoantibodies directly lead to pain or whether pain results from an interaction of antibodies due to the activation of compliment (Guo et al. 2020).

The results of experiments using the PT-TM model suggest that IgM-mediated autoimmunity maybe responsible for early stage CRPS I whereas transferring IgG serum from chronic CRPS I patients to rodents induced the chronic form of the disease (Li et al. 2014). The foregoing descriptions of autoimmune activity are but a small sample of current research studying the relationship of autoimmunity in the etiology of CRPS.

5.5.4   Animal Needlestick Nerve Injury Model (NNI)

Oaklander et al. (2006, 2009) demonstrated that the axon density ipsilateral to the skin of CRPS patients compared to the unaffected contralateral extremity is reduced. This has a bearing on the classification of CRPS I and II in which a known nerve injury, by definition, is associated with CRPS II, whereas neuronal damage unseen and only characterized by ultrastructural nerve biopsy analysis is a fact in CRPS I. Small fibre neuropathy which includes the thinly myelinated A-δ fibres, unmyelinated C fibres and sympathetic axons being targets in this disorder was the stimulus to develop a laboratory method in which reproducible neuronal damage and a percentage injury like that seen in CRPS (29%) can be observed in the animal model (Siegel and Oaklander 2007). An 18-gauge needle was found suitable instrument to induce the desired graduated lesion. Using the tibial nerve as a target, hyperalgesia, edema and abnormal tonic posture developed in the distal tibia of the NNI rodents.

Anne Louise Oaklander

All four animal models described above have already had a dramatic effect on our understanding of the pathogenesis and some potential treatment approaches for this condition. Selection of the most appropriate animal model is fundamental to the anticipated research. With pain in the United States affecting 100 million people and costing upwards of $635 billion per year, to say nothing of the loss of work, a tremendous amount of effort has been devoted to the search for sources of chronic pain, namely, sensitisation of primary sensory neurons and centrally induced neural plasticity as the principal culprits.

5.6　Immune Aspects and CRPS

The circulating and resident immune cells play an important part in this noxious environment. The innate immune response Includes myeloid-lineage cells which in the periphery are inflammatory monocytes and centrally, microglia (Clark et al. 2018). Microglial activation and ATP acting on purinergic-2X (P2X)—fast ligand-gated receptors—and colony stimulating factor-1 (CSF 1)—tissue regulator of macrophages and osteoclasts—in pain states has received significant attention as slow neuromodulators cf. synaptic transmission, over the past 20 years (Xanthos and Sandkühler 2014; Tsuda et al. 2003; Hodge et al. 2011). The toll like receptor 4 (TLR4) which is expressed on myeloid type cells, microglia and neurons has been mooted as one of the key triggers for microglial activation (Lee et al. 2019). The recent elegant study in which the FCIM was used by Huck et al. (2021), found that the TLR 4 receptor has a role in the progression of male and female CRPS, an association that recently many researchers are finding. TLR 4 would seem to be a predominantly male mediator of pain (Sorge et al. 2011). Using knockout mice, were able to ensure that TLR4 is not limited to the microglia. When a TLR 4 antagonist was used, male but not female allodynia was prevented, rather suggesting that central TLR 4 expressing microglia are primarily responsible for this effect? The tibial fracture model of CRPS that was used involves both neuropathic and inflammatory mechanisms. Huck et al. (2021) also noted that TLR4 antagonism used in the foregoing model did improve the intensity of the pain response in both male and female rodents while acknowledging that the mouse strain used in this study might have influenced the extent of the inflammatory component of pain. Huck also noted that in their transgenic mouse model, microglial TR4 contributes to early pain expression in males but less so in females. However, these results have an important bearing on CRPS, in which there is a 4:1 female preponderance (Sandroni et al. 2003). If a low dose of naltrexone a centrally acting opioid TLR 4 antagonist is used, either prophylactically or soon after the onset of acute CRPS I, pain behaviours are abolished, i.e., TLR4 is both time and microglia-dependent for both sexes although recent studies would suggest that chronic pain in females is more dependent on the myeloid-lineage rather than microglial species. While the foregoing data are *etre en train* the do give us a glimpse of how important this receptor for the transition from acute to chronic forms of CRPS.

5.7 Inflammation

A review by Misidou, sees the role of inflammation during the acute phase of CRPS as a primary trigger in its pathophysiology with the expression of pro-inflammatory cytokines such as TNF-α, IL-1β and IL-6 during the acute phase (Misidou and Papagoras 2019). Several meta-analyses and systemic reviews have shown that Il-8 is significantly elevated in the blood during the acute phase as are several other proinflammatory mediators including interferon-γ (IFNγ), IL-2, MCP-1 and bradykinin (Parkitny et al. 2013; Mesaroli et al. 2021). While these mediators are components of the acute stage, they are also found during the cold or chronic phases of CRPS, and as such may respond to anti-inflammatory measures (Urits et al. 2018). Central to the pathophysiology of CRPS, is neurogenic inflammation which following stimulation of nociceptors during trauma, the increase in C-fiber activity results in the secretion of pro-inflammatory neuropeptides SP, CGRP and NK (Weber et al. 2001), Vasoactive substances endothelin-1 (ET-1) and nitric oxide have an important role in early chronic CRPS as do other vasoactive peptides NKB, NPY and VIP on smooth muscle directly impacting the arteriolar blood supply during the acute phase thereby promoting vascular permeability, tissue edema and the extravasation of leukocytes (Groenweg et al. 2006). Other local cells including dendritic and mast cells are activated by CGRP and SP which in turn promote inflammation and sensitisation of Aδ-fibers and CGRP-induced hair growth (Russo et al. 2019; Schlereth and Birklein 2012). While local autonomic nervous system responses including skin colour changes, hyper or hypohidrosis and skin temperature variability these are but only one component of systemic-wide evidence of sympathetic dysfunction reflected by heart rate variability, dysorthostasis amongst other manifestations (Lee et al. 2021; Bartur et al. 2014). During the warm phase, norepinephrine output is reduced and during the chronic phase the α-adrenoceptors are more susceptible to norepinephrine-induced vasoconstriction and SMP. (Teasell and Arnold 2004).

5.7.1 CNS and Autoimmune Responses

The brain's response to CRPS, in particular motor dysfunction involving the primary motor cortex, supplementary motor cortex and posterior parietal cortex has gained a lot of attention during the past three decades (Lee et al. 2022). Reorganisation of the CNS is most likely responsible for the exaggerated reflexes, myoclonic jerks, and dystonia in the ipsilateral extremity that are frequently described (Birklein et al. 2000; di Pietro et al. 2013). Using magnetoencephalography (MEC), Maihöfner evaluated the primary somatosensory cortex representation in an upper limb of 12 CRPS 1 patients who underwent repeated stimulation of the 1–5 digits and the lower lip of both the affected and unaffected extremities with repeated air puffs (Maihöfner et al. 2003). The results demonstrated that the cortical area representing the affected extremity had shifted in the direction of the lip representation (similar topographical cortical changes were reported in amputees with phantom limb pain, Ramachandran and Herstein 1998). Interestingly, in a follow up study, the same investigators found that 10/12 of the previously observed cortical reorganisation had reversed as a result of 1 year's treatment with anti-inflammatory medication and physical therapy, and a reduction of CRPS pain (Maihöfner et al. 2004; Jänig and Baron 2002).

Autoimmunity while suspected for some time as a component of CRPS pathophysiology is supported by studies that have identified the presence of IgG autoantibodies which are agonist to the 2β-adrenergic and muscarinic-2 receptors (Kohr et al. 2011). Autoantibodies have also been found against the α-1a adrenoceptors (Dubuis et al. 2014). The results of a large randomised, double-blind, placebo-controlled crossover trial using low-dose intravenous immunoglobulin (IViG) did not reduce moderately severe CRPS pain in spite of positive evidence from several RTC's, including a smaller controlled study by the same investigators, that IViG can reduce CRPS pain (Goebel et al. 2011, 2017). Although the second study using the same protocol in the larger group of patients did not statistically replicate the earlier results, autoimmunity most likely plays a role in the pathophysiology of CRPS.

5.8 Management of CRPS

The wide range of pathophysiologies and multifactorial nature of these CRPS syndromes presages the diverse methods of treatment, both pharmacologic and interventional that will be necessary for its management. Of necessity, is the restoration of lost function supported by measures that will address (1) the pathophysiology as it is best understood, (2) the management of pain and (3) management of the associated behavioural burden. A multidisciplinary setting is optimal by having the treatment "team" focus on not just physical, occupational and behavioural instruments but by coordinating all elements that may be necessary to manage each individual case. Speed of management and constant observation of a patient's progress is essential if one is to avoid the onset of chronicity, the negative consequences of increasing pain, and the development of deeper connective tissue changes that accrue from disuse (Kessler et al. 2020). Engaging with the patient and their family members is paramount and allows a more active participation of patients in the management of their disease. Patients must be made aware that many treatments will be used because treatment is symptom-directed rather than mechanism-based. During the often-tortuous course of rehabilitation, patient retention is paramount. Any means possible should be used to ensure that each patient is made to feel that they are engaged in their own improvement, the associated clinical personnel being as it were the stewards of the different methods that are used to achieve a good outcome (Neumeister and Romanelli 2020).

5.9 Pharmacologic Measures

Corticosteroids: Their use for acute treatment has already been mentioned in this text. In a recent RCT, Kalita et al. compared prednisone with piroxicam in a small (60) group of stroke patients in which they demonstrated, a significant clinical improvement (Kalita et al. 2016). Opioid's may be necessary to treat pain, certainly during the acute, warm presentation of the syndrome, with the caveat that chronic use is fraught with the development of tolerance, dependency and hyperalgesia, thereby always requiring close monitoring. Their recent incorporation in the "U.K. Guidelines and European Pain Federation Position Statement" for the treatment of early post traumatic CRPS, opioid analgesics and tramadol are recommended to treat trauma-related pain and to facilitate rehabilitative measures. While neuropathic pain can be addressed with anticonvulsants, their analgesic effects are poor, but they can help address sensory changes in the affected extremity. In an RCT, Brown and colleagues compared amitriptyline and gabapentin in 34 children with CRPS, both sleep disturbances and pain relief were achieved, but no interdrug differences were observed (Brown et al. 2016). The response to some vasoactive drugs was described almost four decades ago (Ghostine et al. 1984). Recently, sympatholytic drugs have been used to treat patients with CRPS. The α1 sympathetic blockers phenoxybenzamine or terazosin can reduce pain during the acute stage and have interestingly been equally effective in some longstanding cases. The α_2-adrenergic agonist Clonidine can be effective in a transdermal form to treat hyperalgesia. Bisphosphonates' have received more attention than any other medication that has been used to treat CRPS. However, the mechanism of action is multi-factorial including bone metabolism, mast cell inhibition, NGF regulation, pH modulation (reducing local acidosis) and possibly analgesia. RCT trials support the use of neridronate (Varenna et al. 2013), and pamidronate (Robinson et al. 2004). Calcitonin has demonstrated mixed effects in the treatment of CRPS (O'Connell et al. 2013), and in many respects while achieving some clinical improvement, methodological issues have rather muddied the waters in suggesting that it is a reliable treatment for CRPS. Antioxidants have found a place in the treatment of CRPS ever since van der Laan identified the effect of oxygen radicals in the pathology of the syndrome (Aliminian et al. 2021; Klouchev et al. 2017). DMSO in a cream with ketamine, pentoxifylline and clonidine provided significant pain relief in a high percentage of patients (Russo and Santarelli 2016). A double blind, placebo-controlled study in which DMSO was compared with a placebo found a significant improvement in the pain of CRPS patients after 2 months (Zuurmond et al. 1996). N-acetylcysteine has shown to be effective in dystrophic cases of CRPS (Perez et al. 2003). For prophylaxis, vitamin C is probably effective in the prevention of CRPS in an orthopaedic upper extremity population, although there have been some contradictory publications (Evaniew et al. 2015), its use is recommended on the basis of its negligible side-effect profile and potential efficacy. Botulinum toxin, (BTX-A): BTX-A probably modulates SP and CGRP and sympathetic function (Matak et al. 2017). While BTX-A Can be useful to facilitate rehabilitation of the myofascial syndrome (MFS) and dystonic muscles in patients with upper extremity CRPS, it can also be useful for treating pain in CRPS (Kharkar et al. 2011). Several studies in which BTX-A was used for lumbar sympathetic blocks (LSB) have been described. In a comparative study to evaluate both efficacy and safety, BTX-A and BTX-B were used for lumbar sympathetic blocks. Relief of pain was significant with a greater duration occurring with BTX-B (69 days). No adverse side effects were observed (Lee et al. 2018). As has already been implied, the timing of these pharmacological approaches is param ount to achieving their maximum therapeutic efficacy. Low-dose naltrexone (LDN) is a non-selective pure opioid antagonist with a high affinity for the mu-opioid receptor. LDN binds to the Toll-like receptor 4 where it is a glial antagonist. By interfering with the toll-interleukin receptor (TIR) it can disrupt or prevent the production of inflammatory endproducts, such as (IL)-1, TNF-α, interferon-β and nitric oxide. The "low dose" effect may be less important as an opioid antagonist than its effect in up-regulating endogenous opioid signaling. This is in effect, neuroimmune-modulation (McCusker et al. 2013). Studies have shown that rising levels of endorphin and met-enkephalin (opioid growth factor) stimulate an increase in mu and delta opioids and opioid growth factor receptor levels. A related and beneficial side effect is the neuro-psychological action enhancing the quality of life (Rahn et al. 2011). Several case reports and a recent review describe the successful use of LDN for the management of CRPS (Soin et al. 2021). These results have been echoed by the author over many years in those CRPS patients requiring long term symptomatic medical treatment.

5.10 Interventional Measures

Interventional therapies have long played a role in the treatment of CRPS. Sympathetic blocks remain in many treatment algorithms and while level 2 evidence for their efficacy is limited, they can provide satisfactory analgesia and promote physical function in which case their utility remains as one component in the management of patients with CRPS.

Other regional anesthetic blocks can be useful to provide pain relief for physical therapy. Such practical procedures as brachial plexus analgesia lends itself to an infusion by an implanted catheter using the infraclavicular or axillary approach. Such infusions may be left in situ for 2 weeks, or longer. Epidural analgesia in optimal circumstances can facilitate rehabilitation by an infusion for 6 weeks or longer. Typically, a long-acting local anaesthetic like bupivacaine or ropivacaine together with an opioid (morphine, fentanyl, hydromorphone) analgesic infusion can provide continuous pain relief for physical and occupational therapy management. These techniques require a sterile environment (procedure room, operating room) in which the epidural catheter is introduced under appropriate skin asepsis and subsequent sustainable skin dressing that will allow its management over a prolonged duration by a homecare specialist. The idea is to facilitate the continuum of care without interruption of physiotherapeutic measures in patients either at home or at a specialised rehabilitation facility.

Implantable therapies have become a regular component of many pain treatment programmes. Neuromodulation which includes spinal cord stimulation, peripheral nerve stimulation (SCS), dorsal root ganglion stimulation (DRG), rarely deep brain stimulation (DBS) and Intrathecal drug (IDD) therapy not only have a special place in pain management but also in modification of the pathophysiology that is associated with CRPS. For example, important issues like tissue nutrition can be improved by neurostimulation engendered improvement of blood supply which will also provide favourable conditions in the presence of ischemic vascular disease or ulceration.

5.11   Neuromodulation

Neuromodulation may be indicated when the progress of rehabilitation and the restoration of function is impaired or stalled by increasing pain, motor dysfunction or significant loss of patient compliance. As suggested by the treatment algorithm when simple pharmacologic measures or regional anaesthetic procedures are no longer working, it may be necessary to introduce one of the neuromodulation procedures considered most appropriate for the specific nature of a patient's CRPS. In the past, such modalities were generally considered to be the "last resort" but contemporary practise, based on an increasingly supportive bibliography and currently vast clinical experience now supports their application much earlier for the management of patients with CRPS pain. The results of neuromodulation are salutary (Visnjevac et al. 2017; Chmiela et al. 2021; Van Buyten et al. 2015; Mekhail et al. 2021; Deer et al. 2021). It is customary to undertake a trial stimulation that varies in length depending on the patient characteristics, local regulatory requirements or whether a two-stage procedure that incorporates an initial trial and subsequent surgical implantation of the power supply.

IDD Systems are used generally when there has been a failure of neurostimulation to address pain or motor dysfunction, either the failure of a trial or a subsequent loss of stimulation efficacy. Although morphine is the classical drug and the first to receive FDA approval as an intrathecal agent for use in IDD other opioids are used when a lack of efficacy or side effects preclude the use of morphine. Many different pharmacologic agents have and are being used for intrathecal therapy to improve its efficacy, the local anaesthetic bupivacaine is frequently added and by its action as a sodium channel blocking agent, it has an additive/synergistic effect with the opioid. Clonidine has been extensively used in the treatment of neuropathic pain. Its stability with morphine has been evaluated. In cases of severe neuropathic pain, ziconotide, an n-type voltage-gated Ca-channel blocking agent can replace the use of opioids. While its side effect profile may also preclude its use, it can be extremely effective in responsive patients to manage neuropathic pain. Intrathecal baclofen, a $GABA_B$-receptor agonist a centrally acting muscle relaxant can be used alone in cases of tremor or severe dystonia and is sometimes combined with successful stimulation techniques for symptomatic management of CRPS (Van Hilten et al. 2000; Schwartzman 2000).

William "Bob" van Hilten

Several other drugs have been or are under consideration for intrathecal therapy. The N-methyl-D-aspartate (NMDA) receptor antagonist ketamine has long been suggested (Sator-Katzenschlager et al. 2001; Vranken et al. 2004). Another potential drug is magnesium (Buvanendran et al. 2002). Several preclinical studies have evaluated gabapentin, tizanidine, adenosine, midazolam and tetracaine as possible intrathecal agents. While a trial of neurostimulation is *de rigeur* for all implants, an intrathecal trial of the pharmaceutical agent chosen to meet the specific circumstances for IDD is paramount before surgical implant of the infusion device is contemplated. This may be undertaken in one of two ways, a single intrathecal injection or a continuous infusion via an implanted intrathecal catheter over several days. This latter approach is in my opinion the more appropriate method for a realistic drug trial. In this way, a successful response to the chosen agent for managing a patients' symptoms and at the same time having the ability to observe any side effects replicates the expected response after implantation.

5.12 Rehabilitation

5.12.1 Adults

At the time of diagnosis, treatment will be determined by the nature of the presenting symptoms, their severity, their response to existing or past treatment or whether protracted rehabilitative measures might be indicated. Physical therapy (PT) and occupational therapy (OT) are basic rehabilitation instruments used to achieve functional improvement or resolution of a presenting disability. While Oerlemans found statistically significant changes in the Impairment Sum Score (ISS) temperature side difference hand scores and pain (MPQ) by PT plus Medical Management (MM) compared with either OT plus MM or Social Work (SW), the sample sizes were relatively small to strongly support this approach. While this undoubtedly diluted the element of statistical significance, the trend nevertheless favoured PT (Oerlemans et al. 2000). There is a lack of rigorous controlled trials of different treatment strategies making it necessary for guidance to use consensus reports, reviews and case studies of the various approaches in order to devise a specific rehabilitation program (Stanton-Hicks et al. 2002; Oerlemans et al. 2000). Education strategies are paramount and, in particular the education of patients in the types of procedures that will be used; often overcoming the fear of movement being the most difficult. This latter aspect is the basis for several protocols starting with graded exposure to re-activate movements which can reduce pain related fear and subsequent disability (de Jong et al. 2005) and the use of light touch; desensitisation and pain-adaptive movements; the inclusion of stretching, initially passive but then, introducing the gradual increase in range of movement (ROM) are standard practice (Pleger et al. 2005). These so-called "land" exercises should be incorporated with water therapy and swimming where appropriate (Singh et al. 2004). Stress-loading exercises have a long history (Watson and Carlson 1987) and are particularly effective for upper extremity CRPS; graded axial pressure to the arm and upper arm with so-called "scrubbing" exercises when coupled with stretching are salutory in acute (warm) CRPS. The same principle can be used for lower extremity CRPS. If there is some difficulty during initiation of the exercise program or there is a plateau in progress, mirror image therapy (MVF) can be very helpful; placing a mirror between the affected and unaffected extremity will enable the patient to attempt movement in the affected extremity while visualising (in the mirror) movement in the contralateral extremity (Daly and Bialocerkowski 2009). By playing-back a video of the mirror activity the sensorimotor response will be enhanced to the degree that the patient is encouraged to practise such procedures. Another modality is graded motor imagery (GMI) in which a photographic representation of a hand or foot in various positions is presented to the patient who will in turn identify the laterality of the image (Moseley 2006; Strauss et al. 2020). With training, laterality recognition of the affected limb will improve particularly since the patient is similarly encouraged to extend their practise to the home environment. GMI can materially reduce edema in the fingers of UE CRPS (Moseley 2004). Pain reduction is reduced by up to 20% by GMI (Ferreira et al. 2002). This left right facility will extend to a reorganisation of the premotor cortices. Pain adversely affects cortical activity as seen by fMRI with parietal cortical areas also being activated by actual movement. As a second step, once a patient is asked to visualise a particular posture, in the absence of pain but without actually moving their limb, they will then be asked to imagine movement without pain, i.e., no longer manifesting movement-related pain in the absence of movement—a cortical mechanism. It is now well established that an interdisciplinary pain management program will improve the chances of improving function in patients with CRPS-1 (Singh et al. 2004). Measurement of chronic pain-related cortical activity has considerable value for our future ability to objectively assess pain levels in comparison with subjective measures like VAS scores. Resting state magnetoencephalography (MEG) has allowed us to measure cortical current density and neural connectivity expressed as an amplitude envelope correlation (AEC) on standard fMRI images. This particular study cited both current density and VAS scores that were then calculated for the occipital area and pain-related cortices. Disruption of pain processes and functions in the default mode network does occur in CRPS (Iwatsuki et al. 2021).

From the foregoing description, it should be axiomatic that physical therapy is the primary modality in the management of patients with CRPS; a graded approach with any or all of the foregoing protocols together with whatever supplemental means may be necessary to address severe pain, mention should be made of a recent study that evaluated the safety of pain exposure physical therapy (PEPT) that is used for patients with CRPS (Van de Meent et al. 2011). Although the study number was small—20 patients—the results are revealing and certainly would suggest that similar larger controlled trials should be undertaken. Clearly, PEPT, a progressive-exercise programme is tolerated by CRPS patients. Safety concerns were alleviated and most of the outcome measures related to the International Classification of Functioning (ICF)—activities, personal factors, body functions, and quality of life all improved. The investigators stressed that any local autonomic deregulation or other potential causes for pain are excluded before attempting PEPT. Similarly, arthritis, pseudoarthrosis, osteomyelitis or patients with artificial joint components were also excluded. The principle behind such an approach to rehabilitation is that by managing pain avoidance behaviour, one can restore perhaps the most insidious aspect of CRPS and that is, the maladaptive healing process together with the passive behavioural response to the excessive CRPS pain that is manifested by a withdrawal from physical and social activities and the inevitable spiral towards increasing functional disability development of emotional distress, anxiety, anger and depression (Harden et al. 2004; Stanton-Hicks 2010). Also, as I intimated earlier the authors draw attention to one important factor in the success of PEPT therapy and that is to obviate the misconception frequently held by the medical attendants including physical therapists and physicians, that ongoing pain necessarily implies ongoing tissue damage (De Jong et al. 2005; Vlaeyen et al. 1999). Patients must also be dissuaded from thinking that their pain is a warning signal of continuing damage in their affected region and must therefore be recruited as participants in their own treatment protocol. PEPT can achieve improved function, pain reduction and remission of CRPS without specific medication or other treatments in a significant percentage of patients. As a component of an interdisciplinary treatment programme, it can only add to the overall success by limiting further disability and achieving a high level of successful outcomes. Similar results have been demonstrated in a pain management programme specifically for CRPS in children—below.

5.12.2 Children

Comprehensive rehabilitation should be a given if there is little response to simple therapeutic measures that include PT, OT and recreational therapy (RT) together with medical management for symptom control using anticonvulsants, steroid therapy, tricyclic antidepressants and interventional therapy in the very small number of refractory cases. While pediatric rehabilitation does follow a similar pattern to the management of adult CRPS and the adult algorithm can be used as a guide, it does differ in several important ways. The impact of CRPS on children is emotionally more dramatic than in the adult and the sudden loss of function and level of pain is usually without any historical comparison from the child's perspective. Also, the rapid withdrawal from their peers, school and the influence of their home environment play a much bigger role in their behavioural response to this disease in comparison with how they would react to most other illnesses, than is the case in adult CRPS. These clinical features should be recognized by the treating physician, the response to which requires early introduction of psychological measures, cognitive management and the recognition of any environmental triggers that may aggravate both pain and function (Stanton-Hicks 2010). Graded muscle relaxation using PT and some OT measures as well as play—age related—are important to an early functional return. As mentioned above, the recognition of environmental triggers are particularly relevant and any parental factors such as impending family breakdown, excessive achievement pressure from parent(s) or pre-existing difficulties with peers, teachers, athletic performance at school will alone or together adversely affect the level of pain, function and course of the syndrome in the affected extremity. Depending on the degree of interference that any of the foregoing triggers may have on the response to treatment it may be necessary to separate the patient from this influence, even if it means a short period of inpatient care to remove such factors as the home for instance. As an alternative, a "halfway house" is day hospital rehabilitation that can be very useful in less unresponsive cases.

The following is a description of a 3-week paediatric pain rehabilitation programme at the Cleveland Clinic that has been very successful in managing children with CRPS and other chronic pain conditions (headache, abdominal pain and fibromyalgia) (Banez et al. 2014). It is essentially a 3-week combined inpatient/day hospital programme continued with home rehabilitation and weekly "monitoring" outpatient visits. The programme goals are (1) to help children manage their pain, (2) to restore daily activity, (3) to identify and resolve where possible any environmental factors that are adversely influencing pain and function. By combining inpatient with day hospital components this protocol is unique in comparison with other pediatric programmes that are described in the literature (Eccleston et al. 2003; Hechler et al. 2009; Logan et al. 2012).

The program is designed around 1–2 weeks inpatient care—to obviate any maladaptive parent/child interactions and to facilitate healthy behaviours; to control activity, diet and sleep, using a common treatment philosophy. The third week is a continuation of the interdisciplinary rehabilitation now on a day hospital basis, the child returning home each night. Parents are recruited to assist their child's function and re-entry to their school as indicated.

The treatment plan is individualized and structured in a school format with medical care, psychosocial care, PT, OT, RT and a school program (commensurate with childs' grade). Close communication with the referral source is maintained throughout their treatment. Sub-specialty care is provided by consultation. The treatment protocol is constantly monitored with sub-specialty oversight utilized on an "as needed" basis. Program details are as follows: three individual/family sessions each week; a school program of 1–2 h daily; Aquatic therapy in a 33 °C heated pool focussing on core strengthening, treading, swimming and group games for team building; Leisure, music and recreational aspects occupy I hour each day; to improve parent/child relations, both individual and group sessions, are held each week and each patient has their individual functional and behavioral goals.

A combination of inpatient and day hospital interdisciplinary pain rehabilitation program for children is vindicated by the results of the Banez et al. study (Fig. 5.3).

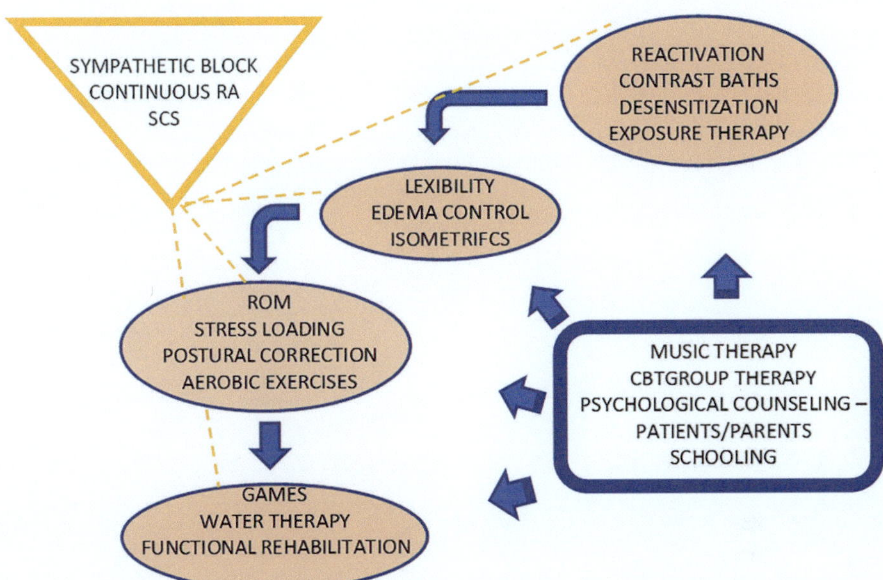

Fig. 5.3 Pediatric functional rehabilitation. The main rehabilitation pathway follows the oval contents from the top-right to the bottom lower left-hand corner of the diagram. Throughout their progress children participate in each of the special treatment groups included in the rectangle. Any divergence, problems with treatment, failure to proceed may call for the additional measures shown in the triangle at any phase of their treatment pathway. *RA* regional anesthesia, *ROM* range of motion, *SCS* spinal cord stimulation

Introduction and Impact of the New Diagnostic Criteria

6.1 Guideline Development

A recurring theme throughout this text has been the absence of any mechanistic basis to substantiate the various pathophysiologies underlying CRPS types 1 & 2. Nevertheless, the Budapest criteria are having a significant influence on clinical practice by improving recognition of the subtle changes that may present in acute onset CRPS and help to distinguish this syndrome from other inflammatory/neuropathic pain syndromes. Identifying later stages or chronic CRPS is less intimidating, the signs are usually well defined with motor/MFPS and possible trophic changes; the pain is invariably mixed neuropathic/nociceptive and allodynia/hyperalgesia either superficial or deep can be severe. By retaining at least one sign in each of the four categories, sensory, vasomotor, sudomotor and motor/trophic of the Budapest criteria has improved the specificity to the point that the differential diagnosis from other conditions that present with similar signs and symptoms is now easier. While no less than five diagnostic criteria have been considered over the past 25 years (Veldman, Bruehl, Orlando Workshop—precursor to the Budapest Workshop—and Ott, Maihöfner CRPS prediction score {CPS}), validation of the Budapest criteria that yielded a sensitivity of .819 and a specificity of .679 provided the impetus for their international adoption. The improved specificity excludes those diseases that have similar clinical features in their differential diagnosis (Ott and Maihöfner 2018). Probably the most significant modification of the Budapest criteria were the results of the validation study (Harden et al. 2010). Unique to these results was the description of two decision rules that increased the usefulness of the criteria, namely, one rule for clinical diagnosis and a second rule to be applied for clinical research. These latter criteria increased the degree of specificity to 0.94 thereby markedly improving the integrity of research data, an example of which are measurements comparing autonomic dysfunction from other signs of neuropathic pain and therefore more likely supporting the diagnosis of CRPS. The authors, Harden et al. did point out a particular limitation of their validation study results which was the inclusion of a very heterogeneous neuropathic pain control group rather than making a comparison between CRPS patients and a much more homogeneous non-CRPS pain patient group that sustained a similar injury. Such criticism was most likely offset by the fact that the heterogeneous control group avoided any bias related to the perhaps unique clinical features associated with a single non-CRPS condition. In spite of the foregoing limitation and the fact study sites were multiple and international, the data support the merit of the Budapest criteria over the original IASP diagnostic criteria as being an instrument that has improved the fidelity of CRPS diagnosis.

In an attempt to standardise the treatment of CRPS, a number of guidelines have been introduced during the past few years an example of which are the standards for diagnosis and its management by **the European Pain Federation Task Force** (Goebel et al. 2019). The importance of these standards is clearly evident in their proposal. Under the following heading's: information and education; pain management; physical and vocational rehabilitation; identifying and treating distress, 17 standards are identified. While much of the detail is described throughout the foregoing text the main point is that such guidance is now increasingly available to medical practitioners in several countries and as a result, the local standard of care will improve. Similar guidelines have been published in the Netherlands, Germany, UK and the US (Grady et al. 2015; Perez et al. 2014; Goebel et al. 2018; Becker et al. 2022; Harden et al. 2022). Algorithms as shown in Fig. 3.2 are still useful in providing a concise thumbnail sketch of a graded rehabilitation pathway aided by specific adjuncts throughout the clinical course any one of which can address specific pain or behavioural issues that may develop during treatment. Such algorithms will be succeeded by much more comprehensive treatment guidelines that address the multitudinous issues that can arise during the treatment. However, implementation of guidelines is not universal because of the variety of clinical practise models, prevailing attitudes and the low prevalence of CRPS limiting an individual practitioner's exposure to clinical cases. Tertiary care medical institutions and large practise groups with an expertise in the management of CRPS become a referral magnet and because of larger caseloads are more likely to adopt their local treatment guidelines. As a corollary to the development of practise guidelines, a recent comprehensive international survey of clinical practise was undertaken by the CRPS service of the National Institutes of Health hospital system in Bath UK (Grieve et al. 2019). The results of this study although somewhat expected are revealing. While the responses represented a large variety of clinical practise, rehabilitation, orthopaedics and pain management made-up the greatest number of medical disciplines. A number of factors describing clinical outcome were seemingly vindicated by the data such as for example, early diagnosis and treatment; while accepting that the temporal nature of this and other disease entities is unique, the data do tend to support better outcomes with early treatment and a bonus of long-term disability prevention. Most patients were seen within 12 month of the onset of their disease and therefore early in their CRPS treatment. These results represent a singular improvement in the outcome of CRPS management in comparison with earlier data (de Mos et al. 2007; Shenker et al. 2015; Lewis et al. 2007). The survey also illustrated particular difficulties experienced by some practitioners such as a lack of resources, a delay in gaining access to early interventions or a fragmentation of available services. Although a smaller number of respondents to the survey used non-standard diagnostic signs and symptoms, most were familiar with the IASP or Budapest diagnostic criteria. Also, a preponderance of respondents acknowledged that an interdisciplinary team approach would achieve the best response. A final and probably just as important an observation as interdisciplinary care, is the collaborative relationship between patients and healthcare professionals reported by the study respondents, is emphasised by the authors of this study. This relationship already referred to in the section on CRPS and Pediatric Management, requires a paternalistic approach involving shared decision-making and therefore more patient-centred care (Treharne et al. 2006; Wertli et al. 2013). Also noted in the survey is the proposition that modern digital technology should be used to expand access to interdisciplinary support for health professionals in those communities having fewer available resources (Cavanagh et al. 2022).

6.2   Tweaking the Terminology

In keeping with the intent and scope of the IASP Special Interest Groups, the SIG on CRPS recently met in Valencia/Spain, September 2019 to review perceived ambiguities in the language of the "Budapest" diagnostic criteria (Fig. 6.1). Foremost amongst these are specific issues related to the clinical and research applications, and the unsatisfactory ICD-11 classification of CRPS as a "focal or segmental autonomic disorder" (ICD-11 BDBA). That this term does not reflect our

Fig. 6.1 Members of the IASP Special Interest Group (SIG) CRPS who met in Valencia, Spain, September 2019:
Front Row: Candace McCabe, Stephen Bruehl, Lone Knudsen, Andreas Goebel, Lorimer Moseley, Normen Hardin, Frank Birklein, Ilona Thomassen, ? , David Clark
Back Row: Jenny Lewis, Janne Gierthmuhlen, Astrid Terkelsen, Christian Maihöfner, Walter Magerl

understanding of CRPS pathophysiology during the past several decades motivated the consensus group to suggest that "chronic primary pain" would be a more appropriate terminology to use and one which would apply to the ICD-11 as the primary *parent* in the classification. The unsatisfactory term CRPS NOS used to codify a clinical case of CRPS that does not completely satisfy the basic diagnostic criteria was addressed by adding a third sub type classified as, "CRPS with remission of some features". These and many other suggestions have now been incorporated in the **"ICD-11 Development Version Text relating to CRPS"**, below. This document is available on the web appendix (http://links.lww.com/PAIN/B358). Re-examination of the terminology is to be expected because at the time of its creation in Orlando, all the participants anticipated that the new terminology would merely be a "work in progress" as its text undergoes modification in response to its widespread application and improved understanding of CRPS pathophysiology. Lastly, perhaps one of the most important aspects of the meeting was the discussion as to whether the distinction between CRPS type 1 and CRPS type 2 should be retained (Stanton-Hicks 2019). Having acknowledged that the only difference between types 1 and 2 is the identifiable nerve lesion, there is no difference between the symptoms and signs of CRPS type 1 and CRPS type 2 and therefore the language in the ICD-11 description should specify that when considering CRPS type 2 as a diagnosis, the signs and symptoms in addition to those neurological signs in the distribution of the injured nerve *must* extend beyond the injured regional nerve territory. The original inclusion of two subtypes in the taxonomy was out of deference to Mitchell's description of causalgia even though at the time it was felt that the two conditions were identical, the injured nerve was considered to be just another concomitant pathology—co-morbidity, an initiating trigger? (Goebel et al. 2021).

6.3   ICD-11 Development Version Text Relating to CRPS— Web Appendix (http://links.lww.com/PAIN/B358)

6.3.1   Description

Complex regional pain syndrome (CRPS) is a chronic pain condition in an extremity with a variable course over time. It is characterized by continuing regional pain (not in a specific nerve territory or dermatome), usually with distal predominance or distal-to-proximal gradient. It typically arises after tissue trauma and is seemingly disproportionate in magnitude or duration to the usual course of pain after such tissue trauma. CRPS is characterized by signs indicating autonomic and neuro-inflammatory changes in the affected body region varying between patients and over time. Often, CRPS is accompanied by significant emotional distress or functional disability. CRPS is multifactorial.

   Complex regional pain syndrome (CRPS) is a chronic pain condition in an extremity (**see note below**) with a variable course over time. It is characterized by continuing (spontaneous and/or evoked) regional pain (not in a specific nerve territory or dermatome), usually with distal predominance or distal- to-proximal gradient. It typically arises after tissue trauma and is seemingly disproportionate in magnitude or duration to the usual course of pain after such tissue trauma.

CRPS is characterized by signs indicating autonomic and neuro-inflammatory changes in the affected body region varying between patients and over time.

Often, CRPS is accompanied by significant emotional distress (anxiety, anger/frustration or depressed mood) and/or functional disability (interference in daily life activities and reduced participation in social roles). As any chronic pain condition, CRPS is multifactorial: biological, psychological and social factors contribute to the disease-phenotype. The diagnosis is based on clinical signs and symptoms alone, so the condition can be diagnosed independent of identified biological or psychological contributors and should only be made if there is no other diagnosis which would better account for the presenting signs and symptoms.

Alternative chronic pain diagnoses to be considered include chronic neuropathic pain and chronic musculoskeletal pain including rheumatological disease. Where regional limb pain is associated with discrete peripheral nerve damage, CRPS can only be diagnosed if CRPS signs and symptoms extend beyond the identified injured nerve territory; clinical features of the nerve lesion (numbness, paranesthesia's) may be restricted to the nerve territory involved; this situation is also termed 'CRPS II'.

Note: Differential diagnosis very important, e.g.: **When this combination of signs and symptoms occurs after mastectomy it is not CRPS?**

6.3.2   Diagnostic Criteria

Criteria: Given the absence of a diagnostic test, CRPS is diagnosed on the basis of clinical signs and symptoms.

Conditions A to E are fulfilled:

Chronic pain in a limb (persistent or recurrent for longer than 3 months) is present

A. The pain is associated with at least one symptom in three of the four following categories*

   Sensory: hyperalgesia and/or allodynia—Conditions A to E are fulfilled:

   A1. *Vasomotor: temperature asymmetry and/or skin color changes and/or skin color asymmetry*

   A2. *Sudomotor/edema: edema and/or sweating changes and/or sweating asymmetry*

   A3. *Motor/trophic: decreased range of motion and/or motor dysfunction (weakness, tremor, dystonia) and/or trophic changes (hair, nails, skin)*

B. *Must display at least one sign at time of evaluation in two or more of the following categories:*

   B1. *Sensory: hyperalgesia (to pinprick) and/or allodynia (for example to light touch, deep somatic pressure, or joint movement)*

   B2. *Vasomotor: temperature asymmetry and/or skin color changes and/or asymmetry*

    B3. *Sudomotor/edema: edema and/or sweating changes and/or sweating asymmetry*

    B4. *Motor/trophic: decreased range of motion and/or motor dysfunction (weakness, tremor, dystonia) and/or trophic changes (hair, nails, skin)*

    B5. *Vasomotor: temperature asymmetry and/or skin color changes and/or skin color asymmetry*

    B6. *Sudomotor/edema: edema and/or sweating changes and/or sweating asymmetry*

    B7. *Motor/trophic: decreased range of motion and/or motor dysfunction (weakness, tremor, dystonia) and/or trophic changes (hair, nails, skin)*

C. *Must display at least one sign at time of evaluation in two or more of the following categories:*

    C1. *Sensory: hyperalgesia (to pinprick) and/or allodynia (for example to light touch, deep somatic pressure, or joint movement)*

    C2. *Vasomotor: temperature asymmetry and/or skin color changes and/or asymmetry*

    C3. *Sudomotor/edema: edema and/or sweating changes and/or sweating asymmetry*

    C4. *Motor/trophic: decreased range of motion and/or motor dysfunction (weakness, tremor, dystonia) and/or trophic changes (hair, nails, skin)*

D. *The pain is associated with at least one of the following:*

    D.1 *Emotional distress due to pain is present.*

    D.2 *Interference with daily life activities and social participation.*

E. *The pain is not better explained by another chronic pain condition.*

    E1. *Vasomotor: temperature asymmetry and/or skin color changes and/or skin color asymmetry*

    E2. *Sudomotor/edema: edema and/or sweating changes and/or sweating asymmetry*

    E3. *Motor/trophic: decreased range of motion and/or motor dysfunction (weakness, tremor, dystonia) and/or trophic changes (hair, nails, skin)*

F. *Must display at least one sign at time of evaluation in two or more of the following categories:*

    F1. *Sensory: hyperalgesia (to pinprick) and/or allodynia (for example to light touch, deep somatic pressure, or joint movement)*

    F2. *Vasomotor: temperature asymmetry and/or skin color changes and/or asymmetry*

    F3. *Sudomotor/edema: edema and/or sweating changes and/or sweating asymmetry*

    F4. *Motor/trophic: decreased range of motion and/or motor dysfunction (weakness, tremor, dystonia) and/or trophic changes (hair, nails, skin)*

    F5. *Vasomotor: temperature asymmetry and/or skin color changes and/or skin color asymmetry*

    F6. *Sudomotor/edema: edema and/or sweating changes and/or sweating asymmetry*

    F7. *Motor/trophic: decreased range of motion and/or motor dysfunction (weakness, tremor, dystonia) and/or trophic changes (hair, nails, skin)*

G. *Must display at least one sign at time of evaluation in two or more of the following categories:*

    G1. *Sensory: hyperalgesia (to pinprick) and/or allodynia (for example to light touch, deep somatic pressure, or joint movement)*

    G2. *Vasomotor: temperature asymmetry and/or skin color changes and/or asymmetry*

    G3. *Sudomotor/edema: edema and/or sweating changes and/or sweating asymmetry*

    G4. *Motor/trophic: decreased range of motion and/or motor dysfunction (weakness, tremor, dystonia) and/or trophic changes (hair, nails, skin)*

    G5. *Vasomotor: temperature asymmetry and/or skin color changes and/or skin color asymmetry*

    G6. *Sudomotor/edema: edema and/or sweating changes and/or sweating asymmetry*

    G7. *Motor/trophic: decreased range of motion and/or motor dysfunction (weakness, tremor, dystonia) and/or trophic changes (hair, nails, skin)*

H. *Must display at least one sign at time of evaluation in two or more of the following categories:*

    H1. *Sensory: hyperalgesia (to pinprick) and/or allodynia (for example to light touch, deep somatic pressure, or joint movement)*

    H2. *Vasomotor: temperature asymmetry and/or skin color changes and/or asymmetry*

    H3. *Sudomotor/edema: edema and/or sweating changes and/or sweating asymmetry*

    H4. *Motor/trophic: decreased range of motion and/or motor dysfunction (weakness, tremor, dystonia) and/or trophic changes (hair, nails, skin)*

    H5. *Vasomotor: temperature asymmetry and/or skin color changes and/or skin color asymmetry*

    H6. *Sudomotor/edema: edema and/or sweating changes and/or sweating asymmetry*

    H7. *Motor/trophic: decreased range of motion and/or motor dysfunction (weakness, tremor, dystonia) and/or trophic changes (hair, nails, skin)*

I. *Must display at least one sign at time of evaluation in two or more of the following categories:*

    I.1 *Sensory: hyperalgesia (to pinprick) and/or allodynia (for example to light touch, deep somatic pressure, or joint movement)*

    I.2 *Vasomotor: temperature asymmetry and/or skin color changes and/or asymmetry*

    I.3 *Sudomotor/edema: edema and/or sweating changes and/or sweating asymmetry*

    I.4 *Motor/trophic: decreased range of motion and/or motor dysfunction (weakness, tremor, dystonia) and/or trophic changes (hair, nails, skin)*

J. *The pain is associated with at least one of the following:*

J1 *Emotional distress due to pain is present.*
J2 *Interference with daily life activities and social participation.*
K. *The pain is not better explained by another chronic pain condition.*

## 6.4	Different types of CRPS can be distinguished

CRPS Type I: Develops after any type of trauma, especially limb fracture or soft tissue lesion. CRPS Type I does not involve discrete nerve damage. Diagnostic signs and symptoms of CRPS Type I and CRPS Type II are identical.

CRPS Type II: Occurs after trauma associated with discrete peripheral nerve damage as indicated by neurological examination, electrodiagnostic testing, or other quasi-objective testing. While clinical features of the nerve lesion (numbness, paresthesia's) are restricted to the nerve territory involved, CRPS signs and symptoms must extend beyond the identified nerve territory.

Diagnostic signs and symptoms of CRPS Type I and CRPS Type II are identical.

CRPS with Remission of Some Features (**NEW** in place of NOS): This category is used to refer to patients who were documented to meet full CRPS criteria (either CRPS 1, or CRPS 2) at an earlier point in time but who currently do not display sufficient signs and symptoms to meet full criteria. Patients in this category are not necessarily improved with regards to pain intensity nor are they free of all CRPS-related signs and symptoms.

Clarification about the application of the diagnostic criteria:

All patients should be asked systematically about all symptoms listed in the criteria at each formal diagnostic evaluation, even if they have not previously reported certain symptoms. CRPS signs and symptoms are clinically observed to fluctuate over time.

– Clarification of the terms "asymmetry" and "changes" as used in the CRPS criteria: For unilateral CRPS, assess asymmetry by comparing the affected side (the side that is most painful) to the unaffected side. For (much rarer) bilateral CRPS, assess changes in the affected limbs relative to an unaffected limb in the patient or to the limbs of a typical healthy individual. Asymmetry is based on clinical judgment only rather than any pre-specified criteria.
– For evaluating possible spreading of CRPS beyond a single limb, the full ICD-11 diagnostic criteria must be applied to each limb individually. True spreading of CRPS is defined as CRPS that meets full ICD-11 diagnostic criteria for multiple limbs—extension of pain to other limbs, which is not unusual, in the absence of other CRPS features is not formally considered to be spreading CRPS.
– **SYMPTOMS** (as given by or elicited from patient, either currently or historically)
  • Hyperalgesia is any minor painful stimulus that is now perceived as intense or prolonged.

- Allodynia is a usually innocuous stimulus now perceived as painful. It is not unusual for a patient to report numbness or hyposensitivity even in the context of clear allodynia in the same body region.
- Temperature asymmetry that is obvious to the patient.
- Color asymmetry that is obvious to the patient.
- Sweating asymmetry that is obvious to the patient.
- Edema that is obvious to the patient (may be described as swelling).
- Dystrophic changes (increased or decreased growth, altered texture) of nails, hair, or skin as described by the patient.
- Motor abnormalities such as weakness, tremor, dystonia (limb locked in unusual position) or jerking of the limb as described by the patient. Decreased range of motion as described by the patient.
- **SIGNS** (as observed by examiner on the exam date)
  - Hyperalgesia is indicated by a mildly noxious stimulus being perceived as intensely painful or the pain lasting longer than the duration of the stimulus. Tested clinically by a single pinprick applied to the affected side (in center of most affected region) and the same site of the contralateral limb
- **Allodynia**: Allodynia can be indicated in multiple ways, all based on reporting pain in response to a normally non-painful stimulus. Stimuli used in clinical assessment can include light touch, vibration, cool or warm temperature, tissue pressure in the affected area, or joint movement.
  - Suggested clinical assessment procedures follow allodynia to light touch as tested by light manual touch (or brush); allodynia to tissue pressure as assessed by pressure applied to a joint or other tissue using the evaluator's finger with just enough pressure to make the fingernail bed of the evaluator blanch (turn white), allodynia to vibration as assessed using a graded tuning fork over bony prominence in affected limb; allodynia to temperature.
  - Temperature asymmetry in the affected area compared to the comparable area on the contralateral extremity that is obvious to the touch of the dorsum of the hand of the examiner.
  - Obvious color asymmetry of a regional nature (i.e., hand, foot, knee or larger region). Please specify the nature of the color changes, for example red, blue, pale, mottled or around a scar.
  - Obvious sweating asymmetry.
  - Asymmetric edema as observed by the examiner or may be quantified by objectively measured volumetry (cm^3) compared side to side, or limb circumference as assessed by tape measure at same location side to side. If a graduated cylinder large enough to hold the extremity or tape measure is unavailable, please just omit the quantification, as long as the categorical "yes/no" response is marked.
  - Dystrophic changes of the affected extremity as noted by the examiner in nails, hair (increased or decreased growth) or skin (shiny, thin, thick etc.) of the affected side. Please describe the nature of these changes.
  - Motor abnormalities as noted by the examiner such as decreased range of motion, weakness, tremor, or dystonia on the affected side.

6.5   Impact of the Terminology

After its introduction in 1995, the new CRPS taxonomy faced many headwinds from the clinical community, in particular those having more than just a brief contact with patients suffering from the syndrome, while its acceptance by the scientific medical community was welcomed because of their greater familiarity with all the arguments supporting this fundamental change of language. Other groups who felt disenfranchised by the change in terminology were patient support categories like the Reflex Sympathetic Dystrophy Association of America (Stanton-Hicks et al. 1995).

Reimbursement issues, particularly in the United States were significantly impacted as the coding of specific interventions used in the management of CRPS and their health insurance coverage were not covered requiring that both the old and new terminologies be used simultaneously when submitting claims for these services. While these particular aspects are no longer of substance, the four-word composition that makes up the terminology is accepted by most for what it is, namely, a bland clinical description of the four clinical descriptors that make up the syndrome until such time as a mechanism is discovered. Those who denigrate the terminology because of its empirical nature and lack of scientific foundation miss the point that its creation was merely a strategy to unify the transfer of and "inform the doctor" of diagnostic information. The fact that exclusion as a component was incorporated in the diagnostic criteria was the need to stress the differential diagnostic tautology of CRPS in order to exclude other conditions such as deep vein thrombosis, Raynaud's disease and diabetes evidence that the terminology does not reflect the signs and symptoms of the syndrome was also a criticism of the terminology. In a manner similar, is the criticism that CRPS induces a nocebo effect, i.e., the signs and symptoms are created by the term itself (Aslaksen and Lyby 2015; Colloca and Benedetti 2007). Several critics over the years, some on more than one occasion have suggested that because no definitive pathophysiology has been identified is another reason that the syndrome does not exist (Borchers and Gershwin 2017; Chang et al. 2019; Ek 2014; Pearce et al. 2005; Verdugo et al. 2010). While many of these criticisms dwell on the lack of a definitive pathophysiology, the very heterogeneity of presenting features over decades has thwarted multiple efforts in cataloguing precise descriptive diagnostic criteria until such time as a mechanistic basis is realised. It is also interesting to note that failure by some scientific observers to acknowledge any connexion between trauma, however small or great, and its biological consequence by phenotypical changes representative of underlying pathology that can be identified during the acute phase and those biomarkers that are found in chronic CRPS is germane to the argument. A recent very important article of this misbelief problem which is as old as medicine itself, under the very eclectic title, "Denying the truth does not change the facts": A systemic analysis of pseudoscientific denial of Complex Regional Pain Syndrome" is a succinct analysis in which Medline, Embase, and Cochrane CENTRAL databases were searched with the help of such key-phrases as: does CRPS exist?; is CRPS a diagnostic entity?; are these claims supported by subsequent arguments? Without delving into detail, the article concludes with the opinion that many of the criticisms ignore the quite abundant CRPS literature—hence evidence base—that critics do not acknowledge the enormous amount of structured and multifaceted nature of research into the disease entity, or even as some authors have suggested, an even higher level of validation then that which is applied to other chronic pain conditions?

The history of medicine has never been kind to those who have made new discoveries or attempted to change the status quo, witness the response to the discovery of Helicobacter pylori (Warren and Marshall 1983). No one in the medical community wanted to believe that a mere bacterium might be responsible for gastritis and or gastric ulceration! The discoverers, a bacteriologist and a pathologist were ridiculed, ostracized, even to the threat of licence removal, and it was only when one of the two discoverers, Barry Marshall drank a solution containing H. Pylori thus inducing gastritis, that turned the tide and the medical community were finally convinced that this discovery was real, momentous and some 20 years later, deserving of a Nobel Prize (2005).

Individual CRPS Case Management   7

Not everyone has access to multidisciplinary care, nor is it necessary for all patients with CRPS to require comprehensive multi-factorial treatment. Most patients presenting with appropriate CRPS diagnostic signs and symptoms will respond to unimodal treatment strategies that will be described in the following case studies that reflect the management of several such cases over a period of 30 years. While accepting the fact that many patients with CRPS do undergo spontaneous remission of their symptomatology, adequate epidemiological studies have still not provided us with a reliable number. What has become obvious from the literature, however, is an increasing number of case reports that describe better outcomes when early treatment and multidisciplinary management is followed. Another factor is an increasing awareness of the benefits of early behavioural measures.

The following descriptions represent a spectrum of CRPS treatment for both adults and children ranging from simple unimodal and physical measures for uncomplicated clinical cases to the more comprehensive management of unresponsive patients often requiring a considerable amount of time and effort or needing a multidisciplinary environment.

7.1 Conventional Medical and Interventional Management

7.1.1 Adult CRPS

Case 1 The following case is quite typical of many patients whose symptoms remit, then recur with or without new trauma, This 50-year-old female teacher's aide presented to the clinic with left upper extremity pain that was exacerbated after she received her annual influenza vaccination. She stated that her symptoms are similar to those she had experienced some 10 years earlier in the right am at which time she sustained a hand injury. She was diagnosed with CRPS which was treated with gabapentin, acetaminophen, and exercise therapy. She is an ardent swimmer. Over a period of 8 months, the condition went into remission, but she continued to

© The Author(s), under exclusive license to Springer Nature Switzerland AG 2024   115
M. Stanton-Hicks, *The Evolution of Complex Regional Pain Syndrome*,
https://doi.org/10.1007/978-3-031-54900-7_7

take gabapentin for a further 12 months. Her new symptoms were described as an episodic burning sensation over the entire left arm that occurs about 15 times a day. She notes that where she had always been able to swim 1/2 mile, she is now unable to do even a quarter of this distance. She had also noticed an increased frequency of dropping objects and the extremity has become cold to touch with increased sweating. Examination revealed weakness at the hand and wrist, the arm was cold, hyperalgesic to pin prick and without sweating on the day of examination. A provisional diagnosis of CRPS secondary to vaccination was entertained. Given the previous success with gabapentin, the patient again received gabapentin 200 mg qid and was advised to continue with her exercise therapy. Duloxetine 60 mg was added and over the succeeding 6 months the patient again went into remission. 10 years later after a motor vehicle accident the same symptoms returned in the left arm and additionally in both lower extremities. The patient also experienced anisocoria of the left eye. Treatment included gabapentin 600 mg qid, duloxetine 60 mg od, and low dose naltrexone (LDN) was titrated to 4.5 mg od. These medications were continued for 18 months and then as symptoms gradually improved they were tapered such that gabapentin 200 mg qid was the only drug required. During this time the patient continued with her exercise programme and resumed her normal daily swimming routine.

7.1.2 Pediatric CRPS

Case 1 The patient is a 12-year-old girl who presented with left foot pain located mostly over the left great toe and ball of her foot. It does not radiate. The pain started when she dropped a table on her left great toe. After 2 weeks the pain returned, more severe and was accompanied by swelling of the foot which was of a reddish-purple discoloration. She was diagnosed and treated for foot sprain but because of continuing symptoms, her mother took her to an orthopaedic surgeon who diagnosed her with CRPS. The patient underwent PT with complete resolution of the pain. One year later, she stubbed the same toe again and underwent further PT/aqua therapy and some psychological counselling with partial resolution. She was then given 20 cycles of hyperbaric oxygen therapy (HBO) following which her symptoms completely resolved for 3 months. These then returned, worsened, and she was given a second course of HBO therapy—20 cycles. She again responded to the treatment, but the improvement was short-lived and only lasted another 3 months.

When she was referred to our clinic, her pain was 7 on the VAS. Examination revealed reduced ROM at the ankle, the foot was blueish, and both foot and ankle were swollen; allodynia was noted over the dorsum of the foot and ankle. Muscle strength was reduced below the knee. A clinical diagnosis of CRPS was confirmed.

Given her past experience and our clinical findings, it was decided that the patient should enter in the 3-week hospital/day-hospital pediatric rehabilitation programme (described earlier in this text). A tunnelled epidural catheter (TEC) was suggested as an aid to therapy, should the patient have any difficulty with therapy during the 3-week programme. This however was never required.

Outpatient medications included Clonidine 0.1 mg od, meloxicam 7.5 mg od, pregabalin 25 mg od, prednisone 10 mg od. and oxycodone/acetaminophen 300 mg/30 mg q6h.prn.

The patient was discharged from the programme completely free of all symptoms and continued with the exercises and the psychological instruments that she had learned during her treatment in the programme. Three years later, the patient was again seen, this time for spontaneous development of CRPS symptoms in the left upper extremity (LUE). The only objective signs noted were coldness over the shoulder and upper arm, weakness of the forearm, hand and wrist and there was decreased movement of the thumb. Some vague patchy pressure-allodynia in the arm was elicited. No edema, temperature side differences or colour changes were found although the patient stated that at times, she had experienced patchy red blotches in the entire extremity. The only medication taken by the patient was ibuprofen prn. She was sent home and prescribed the same therapy she had received after her first visit.

2-years later, she was still having some flare-ups not only in the LUE but also in the LLE. She used many of the strategies she had learned in the 3-weekhospital/day hospital rehabilitation program which helped her to modify or eliminate her symptoms. She continues to attend school, dancing lessons and other leisure activities. She maintains a daily exercise routine. Ibuprofen prn still provides her with reasonable pain relief.

Case 2 This 18-year-old girl presented with a history of having been bucked of a horse 2 years ago. The patient sustained a fractured scaphoid which was repaired/pinned together with external fixation. The pain returned 2 months later and was associated with changes in hair growth, hyperhidrosis in the hand and skin colour and temperature changes. A diagnosis of CRPS resulted in local treatment with stellate ganglion blocks (SGB) which provided excellent pain relief and were repeated at 1-week intervals over a period of 3 months during which time she undertook PT/OT, aquatics therapy and psychological counselling. A complete remission of her symptoms for 3 months was achieved but again returned at which point another SGB gave her another 3 months pain relief and she returned to her normal activities. However, the symptoms and functional impairment of the arm increased. There was also a questionable reinjury from an equestrian incident affecting the same extremity. Her pain described as searing, burning, aching, fluctuating between 5–10/10 VAS and was exacerbated by changes in the weather and or stress. SGB's now only provided about 24 h relief as did a cervical epidural injection. The patient continued to complain of neck, shoulder, and arm symptoms. A cervical MRI showed a C2 mm disc bulge, C6–7 narrowed disc space, but was otherwise normal. Our examination revealed no evidence of CRPS changes at the time. Motor strength and ROM of the entire upper extremity was normal except that increased pain in the forearm, sensitivity and allodynia caused some guarding. In view of the past history and management of a presumed diagnosis of CRPS it was decided to manage the patient in the 3-week inpatient/day hospital program. This was very effective in moderating her symptoms to such a degree that she was able to pass all the specific strategies with which she was presented.

Excerpts from the discharge note are as follows: *"OT noted good functional reach of the LUE but with guarding during the first week. On the second and third weeks, the patient used the arm naturally with all tasks. All sensory band stretching exercises were consistent. The patient verbalized a difference with consistent stretching. PT has total plan for returning to horseback riding and other leisure tasks. Patient will return to school. Will maintain her home exercise program. Education was completed with family on strengthening, aquatics, posture/body mechanics. Endurance training. Patient to be seen again after a month."*

As a sequel to the above description, the patient returned to horseback riding and was attending school. However, a recurrence of symptoms, not only in the extremity, but also the back has necessitated an increase in her pain medications. An MRI of the thoracic spine showed a syrinx. Consultation with a neurosurgeon concluded that no specific treatment would be entertained. She has had to increase her pain medication and is currently working with a therapist who is helping with her home exercise program and is progressing with her schoolwork.

Case 3 In 2002, after a fall at home onto her left hand this 13-year-old girl. Developed. swelling of the fingers. and hand, reduced motion of the wrist, elbow and shoulder together with increased sweating over the forearm and hand. She described her symptoms as burning, sharp, stabbing and throbbing., band-like around her left wrist, which she rated as varying between 5–8 on the VAS. Previous treatment had included. Ultracet, and two courses of steroids which were effective in reducing the swelling. She underwent a SGB which gave her a couple of hours pain relief. Examination revealed a left wrist splint. Marked allodynia to light touch over the hand and dorsum of her forearm was elicited. The fingers. and hand were swollen. Severe limitation of ROM and extreme weakness of elbow and wrist flexion/extension was evident. The diagnosis of CRPS was confirmed.

After discussion with the patient and her parents, an initial T2 sympathetic block was suggested.

Following the procedure, the hand temperature increased from 23.5–34 °C, and the pain reduced from, 8–3/10 VAS. Three months later the patient continued to have symptomatic improvement. Five years later she injured her R thumb and had similar symptoms to those she had previously experienced in her LUE.

Examination revealed a pink hand, allodynia over the hand and wrist, edema of the hand and wrist, and reduced ROM, and weakness. An expression of CRPS in this extremity was the diagnosis. A SGB assisted by ultrasound (US) was considered to be optimal treatment. A temperature change from 22.4–33.4 °C and reduction in the pain scale from. 10/10–6/10 was noted. On this occasion the pain relief was only for 2 days. The decision was taken to place a tunnelled epidural catheter (TEC) and admit the patient to the Pediatric Pain Rehabilitation Unit for comprehensive exercise and behavioural management. This was very successful over a 2-week period, with the exception that a complication related to TEC displacement, required its removal and replacement. It was used for a further 10 days to support PT/OT with significant pain improvement. Supplemental medications during this time were Terazosin 2 mg tid, topiramate 100 mg b.i.d., mirtazapine 15 mg od., tramadol HCL/acetaminophen 37.5/325 mg tab q6h 6h prn.

During the succeeding 2 years, her symptoms which now involved both upper and lower extremities worsened and became refractory to both medications and PT. In 2008, the decision was made to implant a low cervical SCS with coverage available for, both upper and lower extremities. After this procedure, her pain level dropped over 40% in both upper and lower extremities. This has continued to control most of the CRPS pain. Six years later, exhaustion of the IPG needed its replacement. In spite of the need for medical care of other co-morbidities, vis., sacroiliitis, chronic ankle pain, unrelated to her CRPS, she continues to do well, has completed nursing school and is now employed as a nurse. Supportive medications as of 2022 are mexiletine 150 mg bid, terazosin 5 mg od, amitriptyline 10 mg od.

Case 4 In 2000, this 10-year-old presented to our clinic with pain in the right lower extremity (RLE) after she had received a spider bite to the dorsum of her right foot while walking in a zoo. Examination revealed a slightly swollen foot with redness over the dorsum and patchy allodynia over the anterior aspect of the lower leg. Movement was sluggish. When these symptoms persisted and increased in intensity, a provisional diagnosis of CRPS was considered. A Lumbar sympathetic block (LSB) was done with a temperature Δ 22.7–35 °C post-block and the VAS was 3.0. Foot movement returned to normal. Her relief lasted only about 3 weeks and it was determined that she would do well with a continuous epidural infusion (TEC) which would support the prescribed PT and psychologic management in her home-state. Home care services were arranged to manage both the pharmaceuticals, house visits and infusion system. The TEC remained in place for 5 months and was replaced once due to its dislodgement. While her ROM improved with the TEC-facilitated PT, the allodynia and some deep pain remained, her VAS fluctuated between 5–8/10. To address these symptoms, obtain a remission, and given her age, an extended trial of spinal cord stimulation (SCS), not requiring a permanent implanted power supply and not requiring Home Care Services was then tried. For the next 4 months there was a progressive reduction in allodynia and the VAS pain scores hovered around 2–5/10, considerably better than they had been with the TEC. After its removal in May 2004, the patient was maintained on doxepin 10 m od., oxycodone 10 mg od and tiagabine 4 mg hs. for 6 months and then due to sleepiness the tiagabine and doxepin were tapered and stopped. The only medication required during the next 4 years was oxycodone 5 mg od and 10 mg hs. and this was later replaced by tramadol. After becoming pregnant in 2019, the tramadol was discontinued. A delivery under epidural anesthesia was used and interestingly, almost a year post-partum her RLE CRPS symptoms of lesser severity returned. A subsequent LSB was effective in eliminating these, the only residual symptom being a dull ache in the ankle.

7.1.3 Adult Patients and Neuromodulation

Case 1 This 32-year-old male was involved in a motor vehicle accident in which he suffered a severe compound comminuted fracture of his left lower extremity and foot. Within a year (1995), the development of severe neuropathic pain and clinical features consistent with CRPS suggested initial treatment with an SCS would be optimal if he responded with better than 50% symptomatic improvement during trial stimulation. The selected treatment was based on the anticipation that the patient would invariably require many orthopedic procedures in the future.

After a successful trial, a 4-contact laminectomy lead was implanted allowing the patient to undergo successful rehabilitation and a return to his occupation as a legal secretary. Some 3 years later due to loss of adequate pain coverage, it was decided to add a peripheral nerve stimulator (PNS) to the left sciatic nerve to address the lack of adequate coverage. To improve the efficacy of neurostimulation, a trial of terazosin was also found to be effective in reducing the burning sensation and increased sweating in the left lower extremity. Several surgeries including an arthrodesis to correct the varus deformity were unsuccessful in providing a useful weight bearing extremity and the decision to undertake a below knee amputation (BKA) was felt to be appropriate. Three years later, under a continuous epidural anaesthesia the surgery was performed. Within 9 months, the patient developed phantom symptoms in the foot, in addition to which the PNS electrodes were causing symptoms at the stump-prosthetic site necessitating their removal. Medications included pregabalin 150 mg bid and terazosin 10 mg qid. Unfortunately, the pregabalin was discontinued after 6 weeks due to CNS-related side-effects. Twelve years after its implant, it was decided to remove the SCS which was now no longer effective in treating the phantom pain or CRPS symptoms. After his 30-year history of orthopaedic and treatment related interventions, the residual CRPS and phantom pain symptoms, somewhat tempered by time, continue to respond to 10 mg qid of terazosin. The patient has maintained continuous exercise therapy, with weights throughout his illness and has returned to a sedentary occupation requiring the use of a computer.

Case 2 The patient, a 37-year-old white female presented with complaints of right arm pain. She described her pain as being localized over the extensor aspect of the right upper arm with redness noted over the deltoid muscle region. The patient described having a low-grade fever. She had an extensive history of a right scaphoid fracture in 1989 that ultimately required pinning, and she subsequently underwent a wrist fusion in 1990. Based on features that resembled Budapest CRPS diagnostic criteria and regional nature of her symptoms, it was decided to implant a radial nerve PNS to control her pain. While initially successful, decreasing efficacy over 2 years, the PNS was revised in 1992 but the patient never had the same successful coverage that she previously experienced. At the time the patient was undergoing exercise therapy of the right upper extremity, was using oxycodone/acetaminophen for pain control, and continued to work as a legal secretary. Again, after 4 years of less than adequate analgesia, a SCS and rechargeable IPG was implanted after a successful trial. Unfortunately, a wound infection resulted in its surgical removal and re-implantation after 6 weeks of antibiotic treatment. The replacement neuro-stimulator has provided good pain relief for the succeeding 20 years during which time she has received three replacements of her IPG due to power depletion. In spite of the initial technical and subsequent infection-associated implantation complication, neurostimulation in this instance has provided very satisfactory pain relief and allowed the patient to maintain her occupation.

Case 3   The patient a 20-year-old male was referred for evaluation of a right ankle pain and swelling with discoloration. He described a numbness/tingling in the right foot and ankle. Also complained of skin discoloration, swelling, abnormal sweating patterns such as cold sweating and hypersensitivity to touch. The symptoms began 3 years previously after an ankle injury during military training. The symptoms interfere with physical activity, walking and lifting. He described the pain as throbbing, electrical shock-like and rated at eight on the VAS. The pain is constant waxes and wanes reaching a maximum of ten on his worst day. Treatment has included continuous catheter epidural blocks with bupivacaine and clonidine, lumbar sympathetic blocks (7), with a good temperature rise on each occasion. He is currently taking gabapentin 3600 qid, oxycodone/acetaminophen, lidoderm patches, and physical therapy. He states he only has 2 h of uninterrupted sleep at night, and he has trouble concentrating.

Examination revealed allodynia over the entire foot which was dusky and mildly edematous. Movement was sluggish. Numbness was noted over the right medial aspect of the foot. Deep tendon reflexes were 2/5. A diagnosis of CRPS was confirmed.

EMG revealed no nerve lesion. Several MRI's showed edema of the talocalcaneal bones and sub-talar structures. It was decided that the most appropriate treatment at this stage would be a peripheral nerve stimulator, particularly since he lived over 2500 miles away in Alaska. The surgery was performed using a lateral muscle-splitting approach to the sciatic nerve in the thigh. Two-quadrapolar electrodes were placed on the nerve to cover both its divisions. Within 24 h, the patient had good pain control, with a VAS of 5.

Four years later after a soft tissue injury to the affected leg resulted in loss of stimulation over the tibial division, it was decided to explore the PNS nerve site where one of the quadripolar electrodes was found to be displaced 2 cm from the sciatic nerve. This was replaced and good pain control was again re-established. When the patient was seen 5 years later for EOL of the IPG, he was still having good pain control. A new non-rechargeable IPG was replaced. The only medications still required were gabapentin 600 mg od and 1200 mg at night, and amitriptyline 100 mg at night.

Case 4   This 35-year-old right hand dominant female factory worker underwent a carpal tunnel release in 1989 for unremitting pain in the ulnar distribution. Because of unresolved symptoms, the patient then had an ulnar nerve release in 1990 which failed to address her primary symptoms that were complicated by the development of what was then then diagnosed as RSD by her family physician. Although there was little to substantiate this diagnosis in the referring documents, by the patient's own characterization, her symptoms of swelling in the hand, wrist, and forearm, changing skin color and temperature swings in the distal forearm beleid the nature of her disease. She also described extreme weakness (inability to open a jar) and inability to perform her occupation which involved the use of both arms to manage equipment at table height in a production-line. The patient was not taking any medication for her symptoms. After discussion with the patient, it was agreed that an ulnar PNS would be optimal to address her symptoms. The device provided sufficient pain relief to allow her to resume her employment and for the succeeding 7 years when exhaustion of the IPG required its replacement. However, in 2002 due to the development of intermittent stimulation, it became necessary to explore the PNS site. This involved, in addition, neurolysis and release of the antebrachial-cutaneous nerves and implantation of a new IPG electrode. The patient did well until 2019 at which time her IPG was again replaced. The only additional treatment to manage her residual symptoms was occupational therapy. Otherwise, she has continued to work.

7.2   Conclusion

The foregoing cases underscore the capricious nature of CRPS, in particular, the variety and nature of traumatic events that precede its onset, it's multi-faceted presentation and total absence of any stereotypical phenotype. The only significant differences in presenting signs and symptoms are those patients who are seen within hours or days after the onset of clinical CRPS features and patients in whom circumstances, time, or prior treatment is associated with the development of more chronic clinical features. The variety of treatments emphasizes the multiple pathophysiologies that can be active in a temporal sense that require symptomatic management in each case. Also clear from some of the cases is the need to rapidly respond to any attenuation of the treatment. The behavioral aspect requiring treatment is of utmost importance and while in children this is paramount, so are the psychological responses to some distressing aspects of this disease in the adult which also require attention. All of the patients described emphasize the long-term nature of CRPS. While there may be remissions—of symptoms—the preservation of tissue integrity in chronic cases can only be achieved by close attention to the development of any degenerative aspects. Symptomatic pain control can be managed in the majority of cases by quite simple means, some instrumental and some pharmacologic, concierge perhaps, but often lifelong. As noted, some of the patients who responded well to neuromodulation have continued to use their devices in conjunction with their occupation. In such individuals, their particular pathology expressed as regional (extremity) neuropathic pain in most cases, have continued to maintain an active and physically supportive lifestyle. Comorbidity, particularly in association with orthopaedic injuries requires close consultation with orthopaedic and at times, other medical specialties. In regard to PNS, it should be noted that while the revision-rate was unacceptably higher than that for SCS, the only available electrode at the time was the Federal Drug Administration (FDA), SCS-approved device which was never designed for PNS application and therefore is subject to all the vagaries of less-than-optimal neuro-electrode interface and stability with their target nerves. Modern technology has made significant advances in this area such that a future "revision-free" PNS is almost possible.

Some criticism may be directed toward the chronic use of certain drugs described above in the treatment of CRPS symptoms which, either because their age as such have never been subject to adequate randomised statistical studies, or certainly not for their "off label "use in the management of adrenergic phenomena in relation to CRPS (Ghostine et al. 1984). The α-1 postsynaptic adrenergic antagonist and presynaptic α-2 blocking agent, phenoxybenzamine, and terazosin an α-1 adrenergic antagonist, can be highly effective in the management of neuropathic pain and relief of some adrenergic symptoms like sweating. Phenoxybenzamine has been widely used in the treatment of pheochromocytoma and it has immunomodulatory and anti-inflammatory effects (Inchiosa 2013). There is a significant cost difference between phenoxybenzamine and terazosin for which reason terazosin is more commonly prescribed. In many CRPS patients, terazosin has provided relief of adrenergic and neuropathic symptoms as well as improving the efficacy of neurostimulation in such patients.

Opioids including oxycodone, codeine and tramadol have in two of the case histories above, provided prolonged successful analgesia under strictly controlled circumstances in conjunction with ongoing rehabilitation and concomitant schooling or occupation. Such chronic opioid use can be appropriate in conjunction with the overall management strategy in respect to the clinical status at the time.

Also, which should be clear from a few of the pediatric patients discussed, is the prospect that CRPS management may continue well into adulthood. This investment of time and healthcare can be rewarding as most patients will continue to lead active lifestyles and pursue their respective avocation.

Advent of Biomarkers and Their Role in CRPS Diagnosis and Management

8

As a new science, biomarkers have been defined by a **Definitions Working Group (Clin Pharmocol Ther 2001;69:89–95)** in the following language, *"any measurement reflecting an interaction between a biological system and a potential hazard, which may be chemical, physical or biological, the measured response may be functional and physiological, biochemical at the cellular level or a molecular interaction."* In other words, a biomarker can be a surrogate for a phenotypical expression reflected in its measurement, which could be a biochemical or a physiological measurement at a cellular or molecular level. As such, biomarkers can aid in the diagnosis, the results of a therapeutic intervention, or its prognosis and the temporal monitoring of a disease. CRPS, because of its multiple pathophysiologies, all of which are only incompletely understood, lends itself to the use of biomarkers, many of which are already described in the literature. The challenge for the future will be to identify those specific biomarker(s) that can be sourced by practising clinicians and not just in academic medical centres. The following account addresses such questions.

Addition of co-author Frank Birklein, Dr med

8.1   Identification of Biomarkers

Perhaps, the most important discovery will be to find a biomarker(s) that clearly represents a unique characteristic(s) of CRPS pathophysiology (Birklein et al. 2018). The value of skin biomarkers has been recently reviewed by Andronic et al. (2022); 11 studies (299 patients) selected from Cochrane Controlled Trials, MEDLINE, EMBASE and Web of Science Core Collection databases were used. Skin thickness, loss of nerve fibres, mast cell behavior, receptor expression, inflammation, vascular dysregulation, skin hypoxia, and hypersensitivity all of which represent contemporary interests and scope of research into the pathophysiological aspects of CRPS were identified (Morellini et al. 2018; Marinus et al. 2011; Packham and Holly 2018; Oaklander et al. 2006; Lenz et al. 2013; Birklein et al. 2000; Finch et al. 2014; Wesseldijk et al. 2008; Agten et al. 2020; Krämer et al. 2011; Li et al. 2014). In the case of CRPS, *interaction* should include the presenting characteristics of the syndrome, its diagnosis, the response to treatment and whatever temporal changes may occur throughout its duration (Osborne et al. 2015). More than clinical signs and symptoms, biomarkers would yield changes in the underlying pathophysiology that could enable the clinician to make timely adjustments in the course of disease management. Trying to compare an affected limb with the contralateral asymptomatic limb of an affected individual introduces its own bias due to disease predisposition, while this may be de rigueur, far better would be the comparison of CRPS patients with those having other causes of neuropathic pain or even healthy individuals as controls.

Other potential serum biomarkers involve immune activation (Russo et al. 2020), many of which have already been described elsewhere in this text (Kohr et al. 2011; Dubuis et al. 2014). The anti-autonomic immune globulin G (IgG) antibodies are present in both early and late CRPS and already have the potential to identify the syndrome with good specificity.

Mast cells have a great potential as biomarkers since these cells release pruritogenic mediators and algogenic substances both of which are responsible for peripheral sensitisation and underscore vascular leakage and neurogenic inflammation, Gupta et al. 2018. The neurite loss reported by Oaklander and colleagues in 2009 and other investigators suggests a small fibre neuropathy which may be a cause or consequence of CRPS pathophysiology and of course, 1-adrenoceptors have undergone considerable investigation by several investigative teams as potential biomarkers, particularly because of their relationship to the autonomic nervous system and its purported role in the pathophysiology of CRPS 1. Finch et al. 2014, Drummond et al. 2014, Oaklander et al. 2009.

8.2   Current Studies and Classification of Biomarkers

Biomarkers can be characterized into local and systemic types depending on whether they reflect a source of local activity or a systemic response to CRPS. The two main areas of interest are the immune response including inflammatory aspects and neurogenic inflammation.

8.2.1   Local

8.2.1.1 TNF and Inflammatory Cytokines

Already mentioned in the text, the studies by Huygen and his group pioneered the measurement of inflammatory mediators in skin blisters on the arm and as a result, they were not only able to monitor the pro-inflammatory cytokines, TNF-alpha and IL-6, in particular the increase in the affected extremity in response to CRPS, but also their return to normal after treatment with infliximab (Huygen et al. 2004a). In very comprehensive studies in which blister fluid in ipsilateral and contralateral extremities of CRPS patients, both pro-inflammatory and anti-inflammatory cytokines were compared by Lenz et al. (2013). Included were IL-8., IL-124.0, TNF and the chemokine eotaxin. The macrophage inflammatory protein 1β (MIP1β) and MCP1 (Monocyte Chemotactic Protein 1) was also included. Analysis of the data on completion of the study and at 6 months when compared with controls demonstrated a bilateral increase of TNF and MIP1β as well as a decrease of the anti-inflammatory IL-1Rα protein. Side (ipsi vs. contra) differences were not found. Of particular interest is that the characteristics between the increase of pro-inflammatory and decrease in anti-inflammatory cytokines was common to both sides and rather specific for CRPS. Following 6 months of treatment, the levels returned to those seen in control patients. The authors explained the bilaterality of their results as perhaps being a reflection of their experimental design (long suction period of skin blisters—influence of local substances) or a generalised inflammatory reaction (Parkitny et al. 2013). Punch biopsy sampling obviates some of the technique-induced shortcomings of using skin-blister fluid. A comparison of skin TNF levels in osteoarthritis and patients with fractures with CRPS patients (< 6 months duration) were elevated compared with the controls (Krämer et al. 2011). Similarities between TNF levels in fracture healing and CRPS would suggest that skin biopsies might be a useful tool to monitor the underlying pathophysiology of acute and chronic CRPS (Birklein et al. 2015).

8.2.1.2 Mast Cells

Similarly, mast cells and their biologic products obtained by skin biopsy are potential biomarkers (Birklein et al. 2015). Tryptase, a reflection of mast cell activity in the affected extremity has been correlated with the pain score level (Huygen et al. 2004b). As potential biomarkers, these inflammatory mediators offer the potential for diagnosis as well as being a monitor for the response to treatment. The fact that mast cell activity is increased only during the early stage of CRPS is another reason for its fidelity as a diagnostic biomarker (Hermans et al. 2018).

8.2.1.3 Keratinocytes

Keratinocytes are of significant interest due to their proliferation ipsilateral to the affected extremity. Keratinocytes are responsible for epidermal thinning of the skin. This effect is particularly apparent in patients whose disease extends beyond 6 months, the continuing activation of peptidergic nociceptors and their production of neuropeptides, CGRP and substance P (SP) which in turn promotes the production of keratinocytes and the latter also, activation of mast cells. The combined secretion of cytokines by keratinocytes and inflammatory mediators by mast cells sensitises the peptidergic nociceptors, with further activation of keratinocytes—a *circulus viteosis*—and of course, a concomitant increase in pain. Pain itself stimulates the sympathetic nervous system with the production of noradrenaline also a factor in keratinocyte activity. From the foregoing description of inflammatory factors there is ample opportunity to select one or more of these mediators that might represent future biomarkers.

8.2.2   Systemic

Several substances such as T lymphocytes, monocytes, autoantibodies, osteoprotegerin, the α_{1A} and β_2 adrenoceptor proteins, the M_2 muscarinic receptor protein and circulating amino acids and antioxidants, and more recent high mobility molecule group box (HMGB1) to mention but a few that can be sampled from the blood or serum are potential targets that could serve as biomarkers reflecting inflammatory processes in the pathophysiology of CRPS. As is shown in Fig. 8.1 (Birklein and colleagues) the effect of limb immobilisation in the presence of ACE inhibitors that prevent inflammatory mediator cleavage puts the patient at greater risk for CRPS whereas the converse of mobilisation and early anti-inflammatory treatment probably has a prophylactic influence. A conundrum however exists when immobilisation takes precedence after the surgical healing of an unstable fracture. In such circumstances, all efforts are made to mobilise the patient as much as possible while at the same time preserving "site immobility".

8.2.2.1 T-Lymphocytes

Potential biomarkers for CRPS are T lymphocytes (T-cells) of which many subtypes show promise, but many studies have produced conflicting data. Cytotoxic CD8+ have a reduced absolute number of T-cells in comparison with healthy controls, although no change in this subset was found in more recent studies (Ribbers et al. 1998; Ritz et al. 2011). However, in another study, sIL-2R (CD25) a soluble receptor released from T-cells is believed to possess characteristics that could cement its role as a surrogate biomarker for conditions such as lupus erythematosus, rheumatoid arthritis and sarcoidosis (Rubin et al. 1990; Ter Borg et al. 1990; Vorselaars et al. 2015). Most recently in CRPS patients. Significantly raised CD25

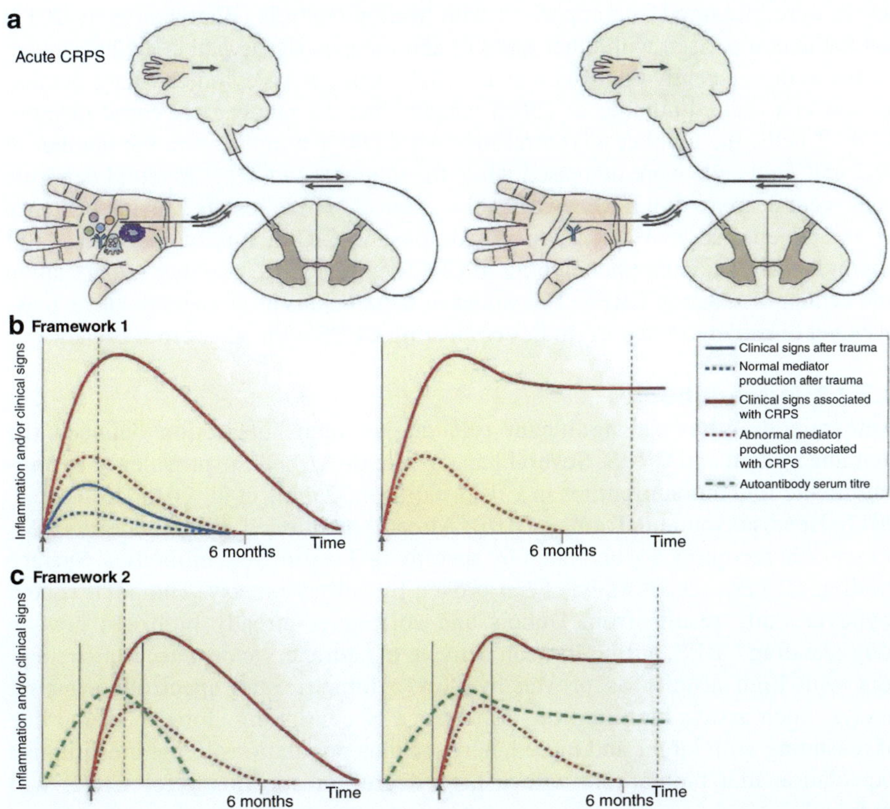

Fig. 8.1 Schematic depiction of acute and persistent Complex Regional Pain Syndrome (CRPS). Reproduced with permission from Birklein F, Ajith SK, Goebel A, Perez RSGM, Sommer C. *Nat Rev Neurol*. 2018;14:272–84. (**a**) Central-pointing arrows represent response to distal excitation. Distal arrows are descending signals representing neurogenic inflammation. (**b**) **Framework 1.a.** Augmented immune activation. Injury response—inflammatory mediators from; mast cells, sensory nerves, keratinocytes, osteocytes (symbols in **a**). **In acute CRPS,** mediator production is temporarily increased triggering augmented post-traumatic clinical signs and primary efferent sensitisation, leading to segmental spinal cord dorsal horn sensitization and sometimes a shift in the cortical representation of the affected limb—abnormal limb perception. Tendency for mediators to normalise around 6 months. **In persistent CRPS,** local inflammatory mediators in the affected limb normalise and CNS changes now drive the clinical picture. CRPS-associated autoantibodies might contribute to some clinical signs. (**c**) **Framework 2: autoantibodies.** Sustained distal limb trauma during a vulnerable time-window characterised by high autoantibody production (solid grey lines), an enhanced trauma-induced inflammatory response renders these antibodies pathogenic and consequently, augmented post-traumatic signs. Autoantibodies might bind to neurons giving rise to afferent sensitization by changes in transduction and transmission properties of these cells or to perineural cells, that then release pronociceptive molecules. Autoantibodies are non-inflammatory—do not activate compliment nor attract immune cells. Acute CRPS normally resolves after cessation of autoantibody production leaving a few cases to sustain the clinical phenotype

levels were measured in comparison with healthy controls (Bharwani et al. 2017) but not in comparison with other types of chronic pain (Bharwani et al. 2021).

It was only recently that Russo et al. (2019). using mass cytometry have demonstrated in a limited number of CRPS patients that the number of central memory $CD8^+$ T cells, the number of central memory $CD4^+$ T lymphocytes, the number of Th1 and Treg cells were increased while the numbers of $CD1c^+$ myeloid dendritic cells were decreased in CRPS vs. healthy controls. These results provide evidence of an antigen-mediated T lymphocyte response in CRPS. Jointly, these results of comprehensive immunophenotyping in CRPS may indicate ongoing inflammation and cellular damage in CRPS. The values of these results as biomarkers have, however, yet to be determined in studies comparing CRPS with other types of pain.

8.2.2.2 Autoantibodies

Autoantibodies have a significant role due to their interaction between the immune system and CRPS. Several papers have described the prevalence of anti-autonomic IgG autoantibodies in CRPS patients (Dubuis et al. 2014; Kohr et al. 2011; Hendrickson and Tormey 2016). Already mentioned, β_2-adrenergic or M_2 muscarinic receptors are activated by specific IgG serum autoantibodies. Surface binding of these receptors has been proven by cell cytometry. The adult rodent myocyte study results from Dubuis and colleagues already highlight how in long-standing CRPS, antibodies can activate α_{1A}-adrenergic or muscarinic receptors with high affinity as proven by flow cytometric and spectrofluorimetric assays. Such assays highlight the importance of continuing along the same line of reasoning with larger and more heterogeneous populations in the presumptive expectation that biomarkers with a high degree of specificity for CRPS will be found.

The use of enzyme-linked immunosorbent assays (ELISA) coated with peptides has a very high specificity and sensitivity in identifying serum autoantibodies for β_2-adrenergic and M_2-muscarinic receptors of CRPS patients. These autoantibodies are in the sub classes IgGI and IgG3 and are without any cross-reactivity (Hendrickson and Tormey 2016). While such observations need to be repeated in large heterogeneous cohorts of CRPS and controls, such autoantibodies have a significant potential for diagnostic biomarkers (Blaes et al. 2004). Very recently, the biological relevance of these CR PS specific autoantibodies for β_2 adrenergic and M_2 muscarinic receptors has been expanded by demonstrating their high affinity binding to, and biological activity on human endothelial cells (Backialakshmi Dharmalingam et al. 2022). The antibodies were cytotoxic for endothelial cells, reduced their proliferation in. In vitro assays. And induced information. Most important, these antibodies were not found in healthy control normal fracture healing controlsa or peripheral arterial occlusive disease patients and were thus potentially useful as biomarkers for CRPS patients. The in vivo passive transfer trauma model (PTTM) and in vitro studies on the primary dorsal root ganglion have been used to identify autoantibodies. Mention has already been made earlier in this text. in which IgG from CRPS patients was injected into rodents all of which in comparison with control animals developed mechanical hyperalgesia and edema, but only in those with injured hind paws (Tekus et al. 2014).

IgM from acute CRPS patients, not from chronic CRPS patients or normal healing fracture patients rekindled hyperalgesia behaviour in fractured mice without B cells (Guo et al. 2020). This finding could help to identify acute CRPS patients in doubtful cases.

8.2.2.3 Monocytes

Monocytes offer significant opportunities as potential biomarkers in CRPS. Three distinct subsets of monocytes are relevant to this discussion, namely $CD14^+CD16^-$, $CD14^+CD16^+$ and $CD14^{+/-}CD16^{++}$ (Ziegler-Heitbrock et al. 2010). In the study by Ritz et al., the abundance of $CD14^+CD16^+$ monocytes found in CRPS correlates with the phenotypic expression of cold allodynia and is very likely responsible for central sensitization—required surrogate biomarkers of the disease (Ritz et al. 2011)? Caution however is required how one interprets whether these monocyte levels reflect a normal inflammatory response or whether in this situation, they may be pathogenic? Also, as a piece of the inflammatory response, circulating monocytes also enter tissues where they develop into macrophages of which there are several subtypes, e.g., the M1 (pro-inflammatory) and M2 (anti-inflammatory) are found in association with CRPS.

8.2.2.4 Micro RNAs

MicroRNAs (miRNA) are small noncoding RNAs that can regulate gene expression. The inhibition of gene expression by miRNAs can be measured in whole blood or in serum-derived small extracellular vesicles (sEVs) which could then be diagnostic or therapeutic biomarkers for CRPS (Wickman et al. 2021). Using the tibia fracture model (TFM) these investigators observed miRNA signatures in serum-derived sEVs from male TFM and control mice. These endosomal sEVs which are between 30 and 150 nm, can transport cell content from a cell over long distances in the body. Such disease pathology changes the miRNA material in the sEVs thereby potentially generating a molecular signature which could be compared against the miRNAs from CRPS patients in whole blood or serum (Orlova et al. 2011). The authors were able to show that differentially expressed miRNAs in the TFM mice shared extensive overlap with miRNAs that are either dysregulated in CRPS patients or treatment. As they emphasized, *"Unifying the role of miRNA with common mechanism in TFM and patients with CRPS can help develop translational approaches to treat CRPS"*. The implications of such studies emphasizes the importance of miRNA profiling using the mouse TFM sEVs to show such a high degree of similarity between miRNA dysregulation and other dysregulation pathways.

In a recent longitudinal study, patients were followed over 2.5 years of standard treatment. They reported lower pain and improved clinical characteristics, 16 patients of which met the criteria for pain relief. Pain relief was associated with stable.

miR-223 expression, whilst pain persistence was characterised by a\ reduction of exosomal miR-233 expression. Thus, progressively reduced miR-233 might be a putative biomarker for chronic CRPS (Reinhold et al. 2022).

The treatment response to ketamine has undergone many studies in which miRNAs have been used to monitor the progress of CRPS. Schwartzman and colleagues have over the years studied the effect of ketamine treatment in patients with CRPS. Of particular interest has been the responder rate, also seemingly gender influenced, is ineffective in over 30% (Schwartzman et al. 2009). Suspecting aberrant gene influence, these investigators have looked at the influence of circulating miRNAs. These substances can inhibit gene expression (binding to the 3′ untranslated region {3'UTR}) of any target miRNA thereby degrading it, resulting in altered gene expression. They have shown that in comparison to healthy controls, miRNAs from CRPS patients classify patients into distinct groups. As such, miRNAs could become therapeutic targets as for example one such miRNA, miR-939 is downregulated in CRPS. This particular mRNA can target a number of proinflammatory genes—VEGFA, TNFα, nNFκB2, IL-6 and NO synthase 2 (NOS2A). Confirmation of this binding is established by reported assays. Over production of miR-939 reduced the concentration of the foregoing genes except NFκB2 which was activated. Thus, these data imply that downregulation of miR-939 in CRPS increases the expression of the foregoing genes, increases inflammatory pain, and generally function at crucial signalling nodes that can worsen the nature of disease (McDonald et al. 2016). An earlier study in females revealed a differential expression of circulating miRNAs in whole blood before and after ketamine treatment. Poor ketamine responders with pain scores <50% decrease had a 22-fold downregulation of mi-605 before their treatment (Douglas et al. 2015). As a sequel to these results, the investigators have explored in which way miR-605 could influence their data. Bioinformatics predict that miR-605 could target epithelial-derived neutrophil-activating peptide 78 (ENA78/CXCL%), in the CXC chemokine family. These peptides can initiate leukocyte migration and initiate an inflammatory cascade (White et al. 2005). Consequently, they have postulated that lower levels of miR-605 may lead to the upregulation of CXCL5 contributing to the inflammation and hyperalgesia in CRPS. In further studies, Douglas and colleagues showed that approximately half a cohort of female patients were responders after blood was drawn, before and 5 days after receiving intravenous ketamine for CRPS. They specifically noted that 33 mi RNAs differed between responders and poor responders. This hinted at the potential reliability of miRNAs to have a predictive value. Of considerable value is the fact that hsa-548d is involved with down regulation in the non-responders, i.e., this miRNA downregulates UDP-glucuronosyltransferase UGT1A1 mRNA. Poor responders have a higher conjugated/unconjugated bilirubin ratio. Vis-a-vis., higher UGT1A1 activity, which negatively impacts ketamine efficacy. Therefore, knowing the miRNA signatures in this case would be of prognostic value for ketamine treatment.

Unravelling how miRNAs can regulate target mRNAs may help to answer how aberrant gene expression can influence CRPS pathology (Birklein et al. 2018a, b). Much of this epigenetic research, while both important and promising of necessity, is carried out in relatively small cohorts and therefore lacks the statistical validation of large numbers.

Another group also pursuing miR-939 targeting of mRNAs found that pro-inflammatory mediators like IL-6 vascular endothelial growth factor, nitric oxide synthase-2 and nuclear factor-κB2 are encoded (McDonald et al. 2016). A genome-wide RNA expression profiling study examined blood from CRPS patients found that of the 80 genes, a fourfold upregulation of the matrix metalloproteinase 9 (MMP9) gene, and many genes were associated with immunity and defence systems. Other studies have not been so successful in achieving statistical significance such as the following in which DNA single-nucleotide polymorphisms (SNPs) were assessed between CRPS and control patients. No differences were seen—even though >200,000 SNPs were analysed. A consequence of a rare disease and the need in genomics to study very large patient cohorts (Janicki et al. 2016)!

Since bone pathology and its increasing turnover in CRPS, particular interest has centred around osteoprotegerin (OPG) a cytokine receptor as a potential biomarker. The receptor activator of nuclear factor-κB, (RANK)/RANK ligand (RANKL)/OPG is a pathway that may clarify the nature of bone symptoms in CRPS. It is an osteoclastogenic inhibitory factor which is present during bone fracture healing and is either a consequence of, or the result of immune system activity. As a member of the TNF family, it is of fundamental interest in the pathophysiology of inflammation in CRPS and a desirable factor for a future biomarker of bone metabolism (Krämer et al. 2014). It should be noted that the type of pain in acute CRPS is primarily a symptom from deep tissues, including bones, and pain hyperalgesia (elicited by pressure) is the most ubiquitous sensory sign (Birklein et al. 2000; Maier et al. 2010). While studies related to OPG. and its place as a biomarker for bone turnover are ongoing, the investigators in this study found increased OPG levels in patients with CRPS, not only associated with fracture, but also in the presence of soft tissue injury suggesting particularly in the latter cases, that inflammatory tissue molecules (cytokines, chemokines etc.) in CRPS pathophysiology upset the interrelationship between osteoblastic and osteoclastic activity. The increase in OPG levels during

CRPS seems to be independent of the acute phase and tend to remain high throughout the disease duration, a useful characteristic for such a biomarker. Validation studies have determined a specificity of 0.8 and a negative predictive value of 84%. These results are comparable with those data in the Budapest research criteria in which such specificity and sensitivities were incorporated. The authors noted some limitations concerning interpretation of the foregoing results included the pilot nature of the study which only used *warm* (acute) CRPS subjects, the comparatively small numbers and the fact that all the control patients were older and tend anyway to have increased OPG levels (Kudlacek et al. 2003).

Epidemiologic data. Suggested that inactivation of the peptide days angiotensin-converting enzyme in patients treated for hypertension increases the odds to develop CR PS after a fracture. A dabsyl-bradykinin (DBK)-based assay was used to investigate very small amounts of serum of patients with CRPS, painful, diabetic neuropathy and healthy controls. The degradation of DBK to fragments L-8 and L-5 was shifted to higher values for DBK L-8 and lower values for DBK L-5 in patients with CRPS after incubation for 1 h. The receiver operating curves recorded values of 0.83 for DBK L-8, indicating a good diagnostic value against healthy controls. Thus, the resolving protease network for mediators like BK might be different in CRPS, and its investigation using the mentioned assay might help in its diagnosis (König et al. 2019). However, some doubts have been raised by the finding that trauma, and not the diagnosis of CRPS, could be a reason. For the disturbance of the serum protease network (König et al. 2022).

8.2.3 Brain (CNS) and Peripheral Biomarkers

A comprehensive study by Jung and colleagues sought to determine how neuroinflammation measured by brain metabolites and peripheral biomarkers in CRPS using Positron Emission Tomography (PET) and Magnetic Resonance Spectroscopy (MRSpec) (Jung et al. 2019).

The central sensitization associated with CRPS that includes NMDA receptors and glial cell induced microglia and astrocyte activation is expressed by pain hypersensitivity, allodynia, pressure hyperalgesia and enhanced temporal summation (Del Valle et al. 2009). Mobile lipids which can be released from cells by hypoxia, necrosis or inflammation are detectable by H-MRS and could provide a window on the neuropathology of neuroinflammation in CRPS. Glial cell activation with the development of gliopathy, a dysregulation of glial function involves both central and peripheral nervous systems. In their previous report, high inflammation levels were found in the basal ganglia and somatosensory cortex, secondary sensory cortices, frontal cortices, insula and the anterior cingulate cortex, in fact, underlying the same regions involved in the neural matrix of CRPS and associated mechanical hyperalgesia (Maihöfner et al. 2005).

Using an activated in vivo marker (activated microglia/brain macrophages)—[11C](R)-PK11195—is a trans locator protein (TSPO)-specific radioligand of neuroinflammation, positive correlations in the distribution volume ratios (DVR) in the right and left insula of CRPS patients but not controls were found, Furthermore using MRS to measure lipid levels, in particular Lip13a and Lip09 relative to total creatinine (tCr) levels, i.e., Lip13a/tCr and Lip09/tCr, a positive correlation with neuroinflammation in the DVR of the left insula of CRPS patients but not in the control patients was found. In addition, pH, glucose, basophil and creatinine in the peripheral biomarkers moved in tandem with either an increase or decrease of inflammation in the microglia. Microglia as mediators of inflammation as such are unique mediators of neuropathic pain (Milligan and Watkins 2009).

As can be seen from the foregoing reports, biomarkers are at the forefront of attempts to identify specific characteristics of CRPS in the pursuit of diagnosis or a response to treatment (Biomarkers 2001). As a component of the multi-pathophysiologic basis of CRPS, inflammation is the target of many potential biomarkers. To be useful, a biomarker should be easy to obtain, measurement of the sample should be possible during routine laboratory testing and obtaining the sample should be minimally invasive in most cases and should be of low risk to the patient. Biomarkers can be selected on the basis of phenotype such as warm type CRPS, or as an inflammatory subtype such as neurogenic inflammation or dysregulation of the immune system (Bruehl et al. 2016; Birklein and Sclereth 2015). While there are no biomarkers for CRPS, a large number of biological molecules have now been identified that have the potential to identify many of the underlying disease processes. Also, multiple risk factors after injury together with the utility of specific biomarkers that can predict the early vs. late stages of the disease entity will be valuable as outcome predictors of specific treatment. The ground has already been laid as potential substances are sequentially identified and supported by statistical analyses. Both the availability and practicality of simple "bed-side" and more esoteric biomarkers of CNS function will obviate the phenotypical characteristics necessary to support the current diagnostic criteria for CRPS. Because CRPS has such a heterogeneous pathophysiology, not one, but multiple biomarkers will of necessity be developed to "fine-tune" its diagnostic accuracy. This area of research not only holds significant promise, but for this particular syndrome and many other diseases, it represents one of the largest and most fertile endeavours to pursue.

Epilogue

This is a story in which the name of a disease entity was changed to harmonize the relationship between its descriptive nature, the fundamental signs and symptoms that are used to make its diagnosis and to reject an older implied mechanistic term on which its original description relied. Another reason for discussing the development of the new terminology is that its final acceptance and promulgation almost coincides with the introduction of the new international classification for diseases ICD-11 which for the first time incorporated chronic pain as a disease entity unto itself (Treede et al. 1992). The IASP Classification of Chronic Pain for the International Classification of Diseases ICD-11. PAIN 2019; 160(1) (ICD-11: BMC Med Inform Decis Mak 2021 Nov 9; 21: an international classification of diseases for the twenty-first century). Classification of medical conditions is no scientific art but rather is an attempt to arrive at some sort of truth alongside of consistency. In most cases, the various categories should be mutually exclusive and at the same time being exhaustive of the data that is to be included. As has been stated by Merskey (1994), "no classification in medicine has ever achieved a single principle, nor can it be expected to do so". While most classifications tend to mirror nature itself, many are artificial and are often convenient as an assembly of facts whereas a phylogenetic classification is superior in all aspects. Classification of an orphan medical syndrome such as RSD with a relatively low prevalence 20.57:100,000 is obviously difficult from the paucity of cases available at any single site as well as the difference in numbers and type of presentation that is inherent between tertiary medical care institutions and private practise. All the more reason however was to begin a process of developing precise criteria which remarkably did achieve considerable success at Schloss Rettershof. However, these diagnostic criteria underwent many iterations as they were worked and reworked during the two intervening consensus workshops in Orlando and Malibu, not forgetting the Research Symposium, surprisingly maintained the basic original framework the literal translation of which are the diagnostic criteria and terminology that was anchored 15 years later in Budapest. The introduction of the concept of animal models by Wilfrid Jänig has transformed our understanding of many pathophysiological processes that are set in motion by CRPS and will continue to broaden our approach to its clinical management in the future (Jänig and Baron 2002, 2003). In the "it's never too late" approach

M. Stanton-Hicks, *The Evolution of Complex Regional Pain Syndrome*, https://doi.org/10.1007/978-3-031-54900-7

for adopting an investigation of the multitudinous inflammatory mechanisms that are set in motion at the outset of the Syndrome, these studies have now been integrated into the vast amount of data that have accumulated over the decades of studying neuropathic and vascular pathology of CRPS. Testimony to contemporary and future investigations tabulated in the appendix is the now widespread interest to better understand and seek a mechanism(s), of which in all probability there are many, will ultimately provide a means of controlling the symptom complex, CRPS. Finding the right biomarker (s) will not only help to improve the clinical diagnosis but will also set the stage for refining those criteria that are essential tools for the diagnostic clinician. I would add that we are now on the cusp of discovering some of these facts.

The rhetorical question is "apart from creating a new terminology what is the real reason to embark on this journey?" the answer is "to provide better patient care based on a better recognition of the pathophysiology that underlies the correct diagnosis and its separation from other medical conditions that may present with similar clinical signs and symptoms, i.e., its differential diagnosis." This is not to say that clinicians should be scientists, but they must at least when making their evaluation, use some scientific principles in support of their clinical hunch and of course, past experience which is the practise of medicine." From the first day—in Dr. Bonica's department—the eternal debate was how is it possible to clinically distinguish this condition, whether named Sudeck's syndrome or reflex sympathetic dystrophy (RSD) from diabetes mellitus, traumatic vasospasm, cellulitis, Raynaud's disease or some other neuropathic state? Also was the fact that while the sympathetic nervous system seemed to be completely anchored in the pathology of RSD why did so many sympathetic blocks fail to either relieve pain or have only a relatively short duration of effect when applied as a means of treatment? Yet at a time when nothing else seemed to help the extreme symptoms with which a patient would present and based on the vast experience in which sympathetic blocks have been used for decades, science entered the discussion in earnest at Schloss Rettershof. This debate provided the underpinnings of the new classification that would set the stage for improved diagnostic criteria and therefore superior tools to replace the older and multiple diagnostic descriptions of RSD at the time for the practising clinician. Eliminating the need for a positive response to a sympathetic block to establish a diagnosis of RSD was a crucial move. Most of all, introduction of the new diagnostic criteria that absolved the need for a diagnostic sympathetic block also provided a clinical terminology that could be universally used by scientists and clinicians alike would enhance communication and the comparison of data. Also, removing the syndrome from the wastebasket of medical conditions and specifically, the inferred behavioural basis under which it has laboured for at least a century has been a triumph of the continuous dialogue throughout each of the consensus workshops. The question of a psychological component as a basis for CRPS has been a recurring theme for ages. The frequently exaggerated presentation of symptoms by these patients if misunderstood by the attending physician is often regarded as neurotic, a not unusual presentation by a lot of chronic pain patients. This has often induced an eristic response cited by a few practitioners who have tried to anchor its causation with psychiatric illness. There is now however an increasing body of literature some of which is adequately blinded, that refutes such a basis for CRPS. Already commented on in the text is the fact that certain personalities—Type A—to symptom magnification, which certainly in some patients might be a factor in their clinical presentation, the majority of patients from all walks of life, blue collar and professions alike describe their new symptoms in similar terms as would any patient with new severe pain, its strange physical expression with a degree of anxiety, at times exaggerated and because of its personal impact, some element of depression. These foregoing aspects support the recommendation that a multidisciplinary environment with the integration of both medical, physical and psychological components will in most cases achieve a fairly good remission. The very illustrative papers by Oerlemans et al. (1999, 2000) of this approach in which the addition of physical therapy interventions with relaxation and cognitive approaches significantly improved pain, range of motion and a general reduction in impairment in comparison to social work alone. This form of management that is so effective in children where in addition play is integrated, is equally effective in an adult population. Until such time as some form of mechanistic basis is found, these interdisciplinary approaches that integrate the foregoing instruments can achieve satisfactory outcomes in a high percentage of patients with CRPS. None of this will be of any value unless a timely diagnosis is made and individual patients are fortunate enough to be treated in such a multidisciplinary setting where speed is of the essence, understanding that delay is a sentence for the development of chronic disease, with all the possibilities of disuse affecting the diseased limb and whole body. While this is a setting for a large number of patients, a few patients acquire their acute symptoms and as is often the case with children, will achieve a remission within days of its onset after a steroid injection, compassionate PT, a block, constant attention and of course, the correct diagnosis. Patients will often require reassurance that their condition is treatable, this against a backdrop of misconceptions, media overload and misinformation together with the negative connotations from relatives and others.

As a proposition for the future, the first chapter on CRPS has been written. Epidemiological studies, prospective clinical and animal research will not only serve as a strategy for further discovery but will also be a blueprint for other medical conditions in which pathophysiological mechanisms remain unknown.

Bibliography

Agten CA, Kobe A, Barnaure I, et al. MRI of complex regional pain syndrome in the foot. Eur J Radiol. 2020;129:109044.

Albrecht PJ, Hines S, Eisenberg E, et al. Pathological alterations of cutaneous innervations and vasculature in affected limbs from patients with complex regional pain syndrome. Pain. 2006;120:244–66.

Alexander FAD. The control of pain. In: Hale DE, editor. Anesthesiology. Philadelphia: Davis; 1954.

Alexander FAD. The genesis of the pain clinic. In: Pain abstracts, Second world congress on pain, vol. 1. Seattle: International Association for the Study of Pain; 1978. p. 250.

Ali Z, Raja SN, Wesselmann U, et al. Intradermal inject of norepinephrine evokes pain in patients with sympathetically maintained pain. Pain. 2000;88:161–8.

Aliminian M, Eraghi A, Chavoshizadah SA, Mohseni M, Mousavi E, Movassaghi SD. Regional vitamin C in Bier block reduces incidence of CRPS-I following distal radius fracture surgery. Eur J Orthop Surg Traumatol. 2021;31:689–93.

Andronic D, Andronic O, Jungle AS, Berli MC, Distler O, Brunner F. Skin biomarkers associated with complex regional pain syndrome (CRPS) Type I: a systemic review. Rheumatol Int. 2022;42:939–47.

Apkarian AV, et al. Cortical responses to thermal pain depend on stimulus size: a functional MRI study. J Neurophysiol. 2000; PMID: 10805705

Arner S. Intravenous phentolamine test: diagnostic and prognostic use in reflex sympathetic dystrophy. Pain. 1991;46:17–22.

Arnold C. Alternative analgesics: new drugs for pain seek to improve on ketamine's benefits. Nat Med. 2017;6:8–10.

Arnold JMO, Teasel RW, MacLeod AP, Brown JE, Carruthers SG. Increased venous alpha-adrenoceptor responsiveness in patients with reflex sympathetic dystrophy. Ann Int Med. 1993;118:619–21.

Aslaksen PM, Lyby PS. Fear of pain potentiates nocebo hyperalgesia. J Pain Res. 2015;8:703–10.

Banez GA, Frazier TW, Wojtowicz AA, Buchannan K, Henty DE, Benore E. Chronic pain in children and adolescents: 24–42 month outcomes of and inpatient day hospital interdisciplinary pain rehabilitation program. J Pediatr Rehabil Med. 2014;7:197–206.

Barowsky EI, Zweig JB, Moskowitz J. Thermal biofeedback in the treatment of symptoms associated with reflex sympathetic dystrophy. J. Child Neurol. 1987;2:229–32.

Baron R, Maier C. Reflex sympathetic dystrophy: skin blood flow, sympathetic vasoconstrictor reflexes and pain before and after surgical sympathectomy. Pain. 1996;67:317–26.

Baron R, Schwarz K, Kleinert A, Schattschneider J, Wasner G. Histamine-induced itch converts into pain in neuropathic hyperalgesia. Neuroreport. 2001;12:3475–8.

Bartur G, Vatine H, Raphaely-Beer N, et al. Heart rate autonomic regulation system at rest and during paced breathing among patients with CRPS as compared to age-matched healthy controls. Pain Med. 2014;15:1569–74.

Basbaum AI, Besson J-M. Towards a new pharmacotherapy of pain. Chichester: Wiley; 1991.

Becker N, Deilmann A, Kowark P, Hildebrand F, Lichte P. [Complex regional pain syndrome—an interdisciplinary view from the surgical consultation]. Chirurgie (Heidelb). 2022;93:819–828.

Bernard C. Introduction to the study of experimental medicine. New York: Collier; 1961.

Bernstein BH. Reflex sympathetic dystrophy in childhood. J Pediatr. 1978;93:211–5.

Betcher AM, Casten DF. Reflex sympathetic dystrophy. Anesthesiology. 1995;16:994–1003.

Betcher AM. Reflex sympathetic dystrophy: criteria for diagnosis and treatment. Anesthesiology. 1955; PMID: 13268914.

Bharwani KD, Dirckx M, Stronks DL, Dik WA, Schreurs MWJ, Huygen FJPM. Elevated plasma levels of sIL-2R in complex regional pain syndrome: a pathologic role for T-lymphocytes? Mediat Inflamm. 2017;2017:2764261.

Bharwani KD, et al. Denying the truth does not change the facts: A systemic analysis of pseudoscientific denial of complex regional pain syndrome. J Pain Res. 2021; PMID: 34737631

Bini G, Hagbarth KE, Hynninen P, Wallin BG. Thermoregulatory and rhythm-generating mechanisms governing the sudomotor and vasoconstrictor outflow in human cutaneous nerves. J Physiol Lond. 1980;306:357–2.

Biomarkers Definitions Working Group. Biomarkers and surrogate endpoints: preferred definitions and conceptual framework. Clin Pharmacol Ther. 2001;57:89–95.

Birklein F, Sclereth T. Complex regional pain syndrome-significant progress in understanding. Pain. 2015;156(Suppl 1):S94–103.

Birklein F, Riedl B, Sieweke N, Weber M, Neundorfer B. Neurological findings in complex regional pain syndromes—analysis of 145 cases. Acta Neurol Scand. 2000;1021:262–9.

Birklein F, On'ell D, Sclereth T. Complex regional pain syndrome: an optimistic perspective. Neurology. 2015;84:89–96.

Birklein F, Ajit SK, Goebel A, Perez RSG, Sommer C. Complex regional pain syndrome-phenotypic characteristics and potential biomarkers. Nat Rev Neurol. 2018;14:272–84.

Birklein F, Ibrahim A, Schlereth T, Kingery WS. The rodent tibia fracture model: a critical review and comparison with the complex regional pain syndrome literature. J Pain. 2018;19(10):1102. e1–1102.e19.

Blaes F, Schmitz K, Tschernatsch M, Kaps M, Krasenbrinl I, Hempelmann G, Brau ME. Autoimmune etiology of complex regional pain syndrome (M. Sudeck). Neurology. 2004;63:1734–6.

Blumberg H. A new clinical approach for diagnosing reflex sympathetic dystrophy. In: Bond MR, Charlton JE, Woolf CJ, editors. Proceedings of the VI th. World Congress on Pain. New York: Elsevier; 1991. p. 399–407.

Blumberg H. Development and therapy of the pain syndrome of reflex sympathetic dystrophy. Clinical expression, experimental investigations, and new pathophysiological considerations. Schmerz. 1988;3:125–43.

Blumberg H, Jänig W. Changes of reflexes in vasoconstrictor neurons supplying the cat hindlimb following chronic nerve lesions: a model for studying mechanisms of reflex sympathetic dystrophy? J Auton Nerv Syst. 1983;7:399–411.

Blumberg H, Jänig W. Clinical manifestations of reflex sympathetic dystrophy and sympathetically maintained pain. In: Wall PD, Melzack R, editors. Textbook of pain. 3rd ed. Edinburgh: Churchill Livingstone; 1994. p. 685–97.

Blumer D, Heilbronn M. Chronic pain as a variant of a depressive disease. The pain- prone disorder. J Nerv Ment Dis. 1982;170:381–406.

Boas RA. Complex regional pain syndromes: symptoms, signs, and differential diagnosis. In: Jänig W, Stanton-Hicks M, editors. Reflex sympathetic dystrophy: a reappraisal: progress in pain research and management, vol. 6. Seattle: IASP Press; 1996. p. 79–92.

Bonica JJ. The management of pain. Philadelphia: Lea and Febiger; 1953.

Bonica JJ. The role of the anesthetist vin the management of intractable pain. Proc R Soc Med. 1954;47:1029–32.

Bonica JJ. Causalgia and other reflex sympathetic dystrophies. Postgrad Med. 1973;53:143–8.

Bonica JJ. Importance of effective pain control. Acta Anaesthesiol Scand Suppl. 1987;85:1–16.

Bonica JJ. Evolution and current status of pain programs. J Pain Symptom Manag. 1990;5:368–74.

Borchers AT, Gershwin ME. The clinical relevance of complex regional pain syndrome type 1: the Emperer's new clothes. Autoimmune Rev. 2017;16:22–3.

Braus DF, Krauss JK, Strobel J. The shoulder-hand syndrome after stroke: a prospective clinical trial. Ann Neurol. 1994;36:728–33.

Brodsky EI, Zweig JB, Moskowitz J. Thermal biofeedback in the treatment of symptoms associated with reflex sympathetic dystrophy. J Child Neurol. 1987;2:229–32.

Brown S, Johnston B, Khush BA, Watkins J, Campbell F, Pehora C, McGrath P. A randomized controlled trial of amitryptiline versus gabapentin for complex regional pain syndrome type I and neuropathic pain in children. Scand J Pain. 2016;13:156–63.

Bruehl S, Carlson CR. Predisposing psychological factors in the development of reflex sympathetic dystrophy. A review of the empirical evidence. Clin J Pain. 1992;8:287–99.

Bruehl S, Lubenow TR, Nath H, Ivankovich O. Validation of thermography in the diagnosis of reflex sympathetic dystrophy. Clin J Pain. 1996;12:316–25.

Bruehl S, Harden RN, Galer BS, et al. External validation of IASP diagnostic criteria for complex regional pain syndrome and proposed research diagnostic criteria. Pain. 1999;81:147–54.

Bruehl S, Galer BS, Harden RN, Saltz SI, Backonja MD, Stanton-Hicks M. Complex regional pain syndrome: are there distinct subtypes and sequential stages of this syndrome? Pain. 2002;95:119–2.

Bruehl S, Maihöfner C, Stanton-Hicks M, Perez R, Vatine JJ, Brunner F, et al. Complex regional pain syndrome: evidence for warm and cold subtypes in a large prospective clinical sample. Pain. 2016;157:1674–81.

Butler SH. Disuse and CRPS. In: Harden RN, Baron R, Jänig, editors. Complex regional pain syndrome: progress in pain research and management, vol. 22. Seattle: IASP Press; 2002. p. 141–50.

Buvanendran A, McCarthy RJ, Kroin JS, Leong W, Perry P, Tuman KJ. Intrathecal magnesium prolongs fentanyl analgesia: a prospective, randomized, controlled trial. Anesth Analg. 2002;95:661–6.

Cavanagh R, Gerson SM, Gleason A, Mackay R, Ciulla R. Competencies needed for behavioral health professionals to integrate digital health technologies into clinical care: a rapid review. J Technol Behav Sci. 2022;17:1–14.

Carron H, Littwiller R. Stellate ganglion block. Anesth Analg. 1975;54:567–70.

Carron H, Weller RW. Treatment of post-traumatic sympathetic dystrophy. In: Advances in neurology. New York: Raven Press; 1974. p. 485–90.

Carter HS. On causalgia and allied painful condition 3: due to lesions of peripheral nerves. J Neurol Psychopathol. 1922;3:1–38.

Chang C, McDonnell F, Gershwin ME. Complex regional pain syndrome—false hopes and miscommunications. Auton Rev. 2019;18:270–8.

Chaplin SR, Bach FW, Pogrel JW, Chung JM, Yaksh TL. Quantitative assessment of tactile allodynia in the rat paw. J Neurosci Methods. 1994;53:55–63.

Chelimsky TCV, Low PAS, Naessens JM, Wilson PR, Amadio PC, O'Brien PC. Value of autonomic testing in reflex sympathetic dystrophy. Mayo Clin Proc. 1995;70:1029–40.

Chen R, Yin C, Hu Q, Liu B, Tai Y, Zheng X, Li Y, Fang J, Liu B. Expression profiling of spinal cord dorsal horn in a rat model of complex regional pain syndrome type-I uncovers potential mechanisms mediating pain and neuroinflammatory responses. J Neuroinflammation. 2020;17:162.

Chmiela MA, Hendrickson M, Hale J, Liang C, Telefus P, Afrin S, Stanton-Hicks M. Direct peripheral nerve stimulation for the treatment of complex regional pain syndrome: a 30-year review. Neuromodulation. 2021;24:971–82.

Christensen K, Jensen EM, Noer I. The reflex sympathetic dystrophy response to treatment with systemic corticosteroids. Acta Chir Scand. 1982;1458:653–5.

Clark D, Tawfik VL, Tajerian M, Kingery WS. Autoinflammatory and autoimmune contributions to complex regional pain syndrome. Mol Pain. 2018;14:1744806918799127.

Coderre TJ, Xanythos DN, Francis L, Bennett GJ. Chronic post-ischemia pain (CPIP): a novel animal model of complex regional pain syndrome-type I (CRPS-I); reflex sympathetic dystrophy produced by prolonged hind paw ischemia and reperfusion in the rat. Pain. 2004;112:94–105.

Colloca L, Finniss D. Nocebo effects, patient-clinician communication, and therapeutic outcomes. JAMA. 2012;307(6):567–8.

Colloca I, Benedetti F. Nocebo hypergalgesis: how anxiety is turned into pain. Curr Opin Anaesthesiol. 2007;20:435–9.

Contrada RJ. Type-A behavior, personality hardiness, and cardiovascular responses to stress. J Pers Soc Psychol. 1989;57:895–903.

Cooper DE, DeLee JC, Ramamurthy S. Reflex sympathetic dystrophy of the knee. Treatment using continuous epidural anesthesia. J Bone Joint Surg Am. 1989;71(3):365–9.

Crabbe JC, Wahlsten D, Dudek BC. Genetics of mouse behavior: interactions with laboratory environment. Science. 1999;284:1670–2.

Daha NA, Banda NK, Roos AR, Beurskens FJ, Bakkerr JM, Daha MR. Complement activation by (auto-) antibodies. Mol Immunol. 2011;14:1656–65.

Daly AE, Bialocerkowski AE. Does evidence support physiotherapy management of adult complex regional pain syndrome Type one? A systematic review. Pain. 2009;13:339–53.

Del Valle L, Schwartzman RJ, Alexander G. Spinal cord histopathological alterations in a patient with longstanding complex regional pain syndrome. Braiin Behav Immun. 2009;23:85–91.

De Jong J, Vlaeyen J, Onghena P, Cuypers C, den Hollander M, Ruijgrtok J. Reduction of pain—related fear in complex regional pain syndrome type one: the application of graded exposure in vivo. Pain. 2005;116:264–75.

de Mos M, de Bruijn A, Huygenv F, Dieleman J, Stricker B, Sturkenboom M. The incidence of complex regional pain syndrome: a population based study. Pain. 2007;129:12–20.

De Takats G. Reflex dystrophy of the extremities. Arch Surg. 1943;46:469–79.

Deer TR, Mekhail N, Provenzano D, et al. The appropriate use of neurostimulation of the spinal cord and peripheral nervous system for the treatment of chronic pain and ischemic diseases: the neuromodulation appropriateness consensus committee. Neuromodulation. 2014;17:515–50.

Deer TR, Esposito MF, Cornidez EG, Okaro U, Fahey ME. Teleprogramming service provides safe and remote stimulation options for patients with DRG-S and SCS implants. J Pain Res. 2021;14:3259–65.

DeLeo D, Magni G, Rossi F, Rosetto F. Psychologische Faktoren beim Sudeck syndrome. Deutsche Medizinische Wockenschrift. 1983;108:7199.

Denmark AS. An example of symptoms resembling tic douleureux produced by a wound in the radial nerve. Med Chir Trans. 1813;4:48–52.

Devor M, Jänig W. Activation of myelinated afferents ending in a neuroma by stimulation of the sympathetic supply in the rat. Neurosci Lett. 1981;24:43–7.

Dharmalingam B, Singh P, Schramm P, Birklein F, Kaps M, Lips KS, Szalay G, Blaes F, Tschernatsch M. Autoantibodies from patients with complex regional pain syndrome induce pro-inflammatory effects and functional disturbances on endothelial cells in vitro. Pain. 2022;163:2445–56.

Di Pietro F, McAuley JH, Parkitny JH, Lotze M, Wand BM, Moseley GL, Stanton TR. Primary somatosensory cortex function complex regional pain syndrome: a systematic review and meta-analysis. J Pain. 2013;14:1001–18.

Diehr P, Diehr G, Koepsell T, et al. Cluster analysis to determine headache types. J Chronic Dis. 1982;35:623–33.

Douglas SR, Shenoda BB, Qureshi RA, Sacan HM, Alexander GM, Perreault M, Barrett JE, Aradillas E, Schwartzman RJ, Ajit SK. Analgesic response ton intravenous ketamine is linked to a circulating MiRNA signature in female patients with complex regional pain syndrome. J Pain. 2015;16:814–24.

Doupe J, Cullen CH, Chance GQ. Post-traumatic pain and the causalgic syndrome. J Neurol Psychiatry. 1944;7:33–48.

Drummond PD, Lancew JW. Clinical diagnosis and computer analysis of headache syndromes. J Neurol Neurosurg Psychiatry. 1984;47:128–33.

Drummond PD, Finch PM, Smythe GA. Reflex sympathetic dystrophy: the significance of differing plasma catecholamine concentrations in affected and unaffected limbs. Brain. 1991;114:2025–36.

Drummond PD, Finch PM, Edvinsson L, Goadsby PJ. Neuropeptide Y in the symptomatic limb of patients with causalgic pain. Clin Auton Res. 1994;4:113–6.

Drummond PD, Finch PM. Sympathetic blockade for complex regional pain syndrome. Pain. 2014;155:2218–2219.

Dubuis E, Thompson V, Leite MI, Blaes F, Maihöfner C, Green Smith D, Vincent A, Shenker N, Kuttikat A, Leuwer M, Goebel A. Longstanding complex regional pain syndrome is associated with activating autoantibodies against alpha-1a adrenoceptors. Pain. 2014;155:2408–17.

Dworkin RH, Backonja M, Rowbotham MC, et al. Advances in neuropathic pain: diagnosis, mechanisms and treatment recommendations. Arch Neurol. 2003;60:1524–34.

Eccleston C, Malleson PN, Clinch J, Connell H, Sourbut C. Chronic pain in adolescents: evaluation of inter-disciplinary cognitive-behavior therapy (CBT). Arch Dus Child. 2003;88:881–5.

Echlin F, Owens FM, Wells WI. Observations on "Major" and "Minor" causalgia. Arch Neurol Psychiatr. 1949;62:183203.

Ecker A. Norepinephrine in reflex sympathetic dystrophy: and hypothesis. Clin J Pain. 1989;5:313–5.

Egle UT, Hoffmann SO, Psychosomatic aspects of reflex sympathetic dystrophy. In Michael Stanton-Hicks, Wilfrid Jänig, Robert Boas, Reflex sympathetic dystrophy, 1990, Boston, Kluwer Academic Publishers. pp. 29.

Ek J. Complex regional pain syndrome type 1does not exist (can it be a normal reaction after immobilization?). BMJ. 2014;348:g2631.

Ekblom A, Hansson P, Lindblom U, Olofsson C. Does a regional nerve block change cutaneous perception threshold outside the anaesthetic area? Implications for the interpretation of diagnostic blocks. Pain. 1992;50:163–7.

Engel GL. "Psychogenic" pain and the pain prone patient. Am J Med. 1959;26:899–918.

Evaniew N, McCarthy C, Kleinlugtenbelt YV, et al. Vitamin C to prevent complex regional pain syndrome in patients with distal radial fractures: a meta-analysis of randomized controlled trials. J Orthop Trauma. 2015;29:e235–41.

Evans JA. Reflex sympathetic dystrophy. Surg Clin North Am. 1946;26:78–90.

Evans JA. Reflex sympathetic dystrophy. Surg Clin North Am. 1946b;26:78–90.

Fleming WW, Trendelenburg U. The development of supersensitivity to norepinephrine after pretreatment with reserpine. J. Pharmacol Exp Ther. 1961;133:41–51.

Ferreira M, Ferreira P, Latimer J, Herbert R, Maher C. Does spinal manipulative therapy help people with chronic low back pain? Aust J Physiother. 2002;48:277–84.

Fields HL, Basbaum A. Central nervous system mechanisms of pain modulation. In: Wall PD, Melzack R (Eds). Texbook of Pain. 4th ed. Edinburgh: Churchill Livingstone. 1999. pp. 309–29.

Finch PM, Drummond ES, Dawson LF, Phillips JK, Drummond PD. Up-regulation of cutaneous alpha I-adrenoceptors in complex regional pain syndrome type I. Pain Med. 2014;14:1945–56.

Flor H, Fydrich T, Turk DC. Efficacy of multidisciplinary pain treatment centers: a met-analytic review. Pain. 1992;49:221–30.

Fordyce WE, Fowler RS, Lehman JF, Delateur BJ, Sand PL, Trieschmann RB. Operant conditioning in the treatment of chronic pain. Arch Phys Med Rehabil. 1973;54:399–408.

Fowler CJ, Carroll MB, Burns D, Howe N, Robinson KJ. A portable system for measuring cutaneous thresholds for warming and cooling. Neurol Neurosurg Psychiatry. 1987;50:1211–5.

Franz DN, Perry RS. Mechanisms for differential block among single myelinated and non-myelinated axons by procaine. J Physiol. 1974;236:193–210.

Galer BS, Bruehl S, Harden RN. IASP diagnostic criteria for complex regional pain syndrome: a preliminary empirical validation study. International Association for the Study of Pain. Clin J Pain. 1998a;14(1):48–54.

Galer BDS Geertzen JH, Dijkstra PU, Groothoff JW, ten Duis HJ, Eisma WH. Reflex sympathetic dystrophy of the upper extremity—a 5.5-year follow-up Part 11. Social life events, general health and changes in occupation. Acta Orthop Scan Suppl. 1998b;279:19–23

Galer BS, Butler S, Jensen M. Case reports and hypothesis: a neglect-like syndrome may be responsible for the motor disturbance in. reflex sympathetic dystrophy. J Pain Symptom Manag. 1995;10:358–92.

Gasser HS, Erlanger J. The role off fiber size in the establishment of a nerve block by pressure or cocaine. Am J Phys. 1929;85:581–91.

Geertzen JH, de Bruijn H, de Bruijn-Kofman AT, Arendzen JH. Reflex sympathetic dystrophy: early treatment and psychological aspects. Arch Phys Med Rehabil. 1994;75:442–6.

Geertzen JH, Dijkstra PU, Groothoff JW, ten Duis HJ, Eisma WH. Reflex sympathetic dystrophy of the upper extremity—a 5.5-year follow-up. Part I. Impairments and perceived disability. Acta Orthop Scand Suppl. 1998;279:12–8.

Geha PY, Baliki MN, Harden RN, Bauer WR, Parrish TB, Apkarian AV. The brain in chronic CRPS pain: abnormal gray-white matter interactions in emotional and autonomic regions. Neuron. 2008;60:570–81.

Gershon E, Wu TX, Shir Y, et al. Inheritance of phantom limb pain in Israeli Jewish veteran amputees is associated with a locus on chromosome 6P21. Neurosci Lett. 1999;52(Suppl):S110.

Ghostine SY, Comair YG, Turner DM, Kasell NF, Azar CG. Phenoxybenzamine in the treatment of causalgia. Report of 40 cases. J Neurosurg. 1984;60:1263–8.

Gibbons JJ, Wilson PR. RSD score: criteria for the diagnosis of reflex sympathetic dystrophy and causalgia. Clin J Pain. 1992;8:260–3.

Gissen AJ, Covino BG, Gregus J. Differential sensitivities of mammalian nerve fibers to local anesthetic agents. Anesthesiology. 1980;53:467–74.

Goebel A. Morphine and memantine treatment of long-standing complex regional pain syndrome. Pain Med. 2012;13:357–8.

Goebel A, Blaes F. Complex regional pain syndrome, prototype of a novel kind of autoimmune disease. Autoimmun Rev. 2013;12:682–6.

Goebel A, Leite MI, Yang L, Deacon R, Cenden CM, Fox-Lewis A, Vincent A. The passive transfer of immunoglobulin G serum antibodies from patients with longstanding complex regional pain syndrome. Eur J Pain. 2011;15:504.e1–6.

Goebel A, Bisla J, Carganillo R, Cole C, Gupta F, Kelly JM, McCabe C, Milligan H, Padfield MC, Phillips C, Poole H, Saunders M, Serpell M, Shenker N, Shoukrey K, Wyatt L, Ambler G. A randomized placebo-controlled phase III multicenter trial: low-dose intravenous immunoglobulin treatment for long-standing complex regional pain syndrome (LIPS trial). Southampton: NIHR Journals Library; 2017.

Goebel A, Callaghan T, Jones S. Patient consultation about a trial of therapeutic plasma exchange for complex, regional pain syndrome. J Clin Apher. 2018;33:661–65.

Goebel A, Barker C, Birklein F, Brunner F, Casale R, Eccleston C, Eisenberg E, NcCabe CS, Moseley GL, Perez R, Perrot S, Terkelsen A, Thomassen I, Zyluk A, Wells C. Standards for the diagnosis and management of complex regional pain syndrome: results of a European Pain Federation task force. Eur J Pain. 2019;23:641–51.

Goebel A, Birklein F, Brunner F, Clark JD, Gierthmühlen J, Harden N, Huygen F, Knudsen L, McCabe C, Lewis J, Maihöfner C, Magerl W, Moseley GL, Terkelsen A, Thomasson I, Bruehl S. The Valencia consensus-based adaptation of the IASP complex regional pain syndrome diagnostic criteria. Pain. 2021;162:2346–8.

Goldsmith DP, Vivino FB, Eichenfield AH, Athreya BH, Heyman S. Nuclear imaging and clinical features of childhood reflex neurovascular dystrophy: comparison with adults. Arthritis Rheum. 1989;32:480–5.

Gonzales R, Sherbourne CD, Goldyne ME, Levine JD. Noradrenaline-induced prostaglandin production by sympathetic postganglionic neurons is mediated by α_{2-} adrenergic receptors. J Neurochem. 1991;57:1145–50.

Goris RJA. Treatment of reflex sympathetic dystrophy with hydroxyl radical scavengers. Unfalchir. 1985;88:330–2.

Goris RJA, Dongen LM, Winters MA. Are toxic radicles involved in the pathogenesis of reflex sympathetic dystrophy. Free Radic Res Comm. 1987;3:13–8.

Grachev ID, Fredrickson BE, Apkarian VA. Abnormal brain chemistry in chronic back pain: an in vivo proton magnetic resonance spectroscopy study. Pain. 2000;89:7–18.

Grady K, et al. Core standards for pain management services in the UK. London: Faculty of Pain Medicine Royal College of Anaesthetists; n.d.

Grieve S, Llewellyn A, Jones L, Manns S, Glanville V, McCabe CS. Complex regional pain syndrome: an international survey of clinical practice. Eur J Pain. 2019;23:1800–903.

Groenweg JG, Huygen FJPM, Heijmans-Antonissen C, Niehof S, Zijlstra FJ. Increased endothelin-1 and diminished nitric oxide levels I blister fluids of patients with intermediate cold type complex regional pain syndrome type 1. BMC Musculoskelet Disord. 2006;7:91.

Guo TZ, Wei T, Huang TT, Kingery WS, Clark JD. Oxidative stress contributes to fracture/cast-induced inflammation and pain in a rat model, of complex regional pain syndrome. J Pain. 2018;19:1147–56.

Guo T-Z, Wei T, Tajerian M, Clark JD, Birklein F, Goebel A, Li W-W, Sahbaie P, Escolano FL, Herrnberger M, Kramer HH, Kingery WS. Complex regional pain syndrome patient IgM has pronociceptive effects in the skin and spinal cord of tibia fracture mice. Pain. 2020;16:797–809.

Gupta K, Harvimal T. Mast cell-neural interactions contribute to pain and itch. Immunol Rev. 2018;282:168–87.

Guzman J, Esmail R, Karjalainen K, et al. Multidisciplinary rehabilitation for chronic low back pain: systemic review. Br Med J. 2001;322:1511–6.

Hansson P. Possibilities and potential pitfalls of combined bedside and quantitative somatosensory analysis of pain patients. In: Bovie J, Hansson P, Lindblom U, editors. Touch, temperature and pain in health and disease: mechanisms and assessments, progress in pain research and management, vol. 3. Seattle: IASP Press; 1994. p. 113–32.

Harden RN, Bruehl S, Galer BS, et al. Complex regional pain syndrome: are the IASP diagnostic criteria valid and sufficiently comprehensive? Pain. 1999;83:211–9.

Harden RN, Rudin NJ, Bruehl S, Kee W, Parikh DK, Kooch J, Duc T, Gracely RH. Increased systemic catecholamines in complex regional pain syndrome and relationship to psychological factors: a pilot study. Anesth Analg. 2004;99:1478–85.

Harden RN, Bruehl S, Stanton-Hicks M, Wilson PR. Proposed new diagnostic criteria for complex regional pain syndrome. Pain Med. 2007;8:326–31.

Harden RN, Bruehl S, Perez RSGM, Birklein F, Marinus J, Maihöfner C, Lubenow T, Buvanendran A, Mackay S, Graciosa JJ, Mogilevski M, Ramsden C, Chont M, Vatine J-J. Validation of proposed diagnostic criteria (the "Budapest Criteria") for complex regional pain syndrome. Pain. 2010;150:268–74.

Harden RN, McCabe CS, Goebel A,. Massey M, Suvar T, Grieve S, Bruehl S. Complex regional pain syndrome: Practical diagnosis and treatment guidelines, 5th edition. Pain 2022;10:10–23

Hechler T, Dobe M, Kosfelder J, Damschen U, Hubner B, Blankenburg M, Sauer C, Zeernikow B. Effectiveness of a 3-week multimodal inpatient pain treatment for adolescents suffering from chronic pain. Clin J Pain. 2009;25:156–66.

Helyues Z, Tekus V, Szentes N, Pohoczky K, Botz B, Kiss T, Kemeny A, Környei Z, Toth K, Lenart N, Abraham H, Pinteaux E, Francis S, Denes A, Goebel A. Transfer of complex regional pain syndrome to mice via human antibodies is mediated by interleukin-1-induced mechanisms. Proc Natl Acad Sci U S A. 2019;116:13067–76.

Hemler DE, Mc Auley RA, Belandres PV. Common clinical presentations among active-duty personnel with traumatically induced reflex sympathetic dystrophy. Mil Med. 1988;153:493–5.

Hendrickson JE, Tormey CA. Red blood cell antibodies in hematology/oncology patients: interpretation of immunohematologic tests and clinical significance of detected antibodies. Hematol Oncol Clin North Am. 2016;3:635–51.

Hermans MAW, Schrijver B, van Holten-Neelen C, Girth Van Eijk R, Hagen PM, van Daele PKLA, et al. The JAK1/JAK2 inhibitor ruxolitinib inhibits mast cell degranulation and cytokine release. Clin Exp Allergy. 2018;48:1412–20.

Hodge JM, Collier FM, Pavlos NJ, Kirkland MA, Nicholsom GC. M-CSF potently augments RANKL-induced resorption activation in mature human osteoclasts. PLoS One. 2011;6:e21462.

Hord AH, Rooks MD, Stephens BO, et al. Intravenous bretylium and lidocaine for treatment of reflex sympathetic dystrophy: a randomized, double-blind study. Anesth Analg. 1992;7:818–21.

Hübner L. Das Sudeck-syndrom. Der Landartzt. 1957;3:651–7.

Huck NA, Silezar-Doyle J, Haight ES, Ishida R, Forman TE, Wu S, Shen H. Transition to chronic pain: a focus on sex differences. J Neurosci. 2021;41:4349–65.

Huygen FJP, Niehof S, Ziljstra FJ, van Hagen PM, van Daele. Successful treatment of CRPS I with anti-TNF. J Pain Symptom Manage. 2004a;27:101–3.

Huygen FJ, Ramdhani N, Van Toorenenbergen AS, Klein J, Zijlstra FJ. Mast cells are involved in inflammatory reactions during complex regional pain syndrome type 1. Immunol Lett. 2004b;91 (2–3):147–54.

Inchiosa R. MA. Phenoxybenzamine in complex regional pain syndrome: potential role and novel mechanisms. Anesthesiol Res Pract 2013;2013:: 9786156/2013/978615 Epub 2013. PMID 24454356.

Iwatsuki K, Hoshiyama M, Yoshida A, Uemura JI, Hoshino AS, Morikawa I, Nakagawa Y, Hirata H. Chronic pain-related cortical neural activity in patients with complex regional pain syndrome. IBRO Neurosci Rep. 2021;13:208–15.

Jaennerod M, Arbib MA, Rizzolatti G, Sakata H. Grasping objects: the cortical mechanisms of visuomotor transformation. Trends Neurosci 1995;18:314–20.

Janicki PK, Alexander GM, Eckert J, Postula M, Schwartzman RJ. Analysis of common single nucleotide polymorphisms in complex regional pain syndrome: genome wide association study approach and pooled strategy. Pain Med. 2016;17:2344–52.

Janicki TI. Chronic pelvic pain as a form of chronic regional pain syndrome. Clin Obst Gyn. 2003;46:797–803.

Jänig W. Causalgia and reflex sympathetic dystrophy: in which way is the sympathetic nervous system involved? Trends Neurosci. 1985;8:471–7.

Jänig W. Neuronal mechanisms of pain with special emphasis on visceral and deep somatic pain. Acta Neurochir Suppl (Wein). 1987;38:15–32.

Jänig W. The autonomic nervous system. In: Schmidt FRF, Thews G, editors. Human physiology. Berlin: Springer; 1988.

Jänig W. Guest editorial: experimental approach to reflex sympathetic dystrophy and related disorders. Pain. 1991;46:241–5.

Jänig W and Koltzenburg M. What is the interaction between the sympathetic terminal and the primary afferent fiber? In AI Basbaum and JM Besson (Eds), Towards a New Pharmacotherapy of Pain, Dahlem Workshop Reports, John Wiley & Sons, Chichester. 1991:331–352

Jänig W. The sympathetic nervous system in pain. Eur J Anaesthesiol Suppl. 1995;10:53–60.

Jänig W, Baron R. Complex regional pain syndrome is a disease of the central nervous system. Clin Auton Res. 2002;12:150–64.

Jänig W, Baron R. Complex regional pain syndrome: mystery explained? Lancet Neurol. 2003;2:687–97.

Jänig W, Stanton-Hicks M. Reflex sympathetic dystrophy: changing concepts and taxonomy. Pain. 1995;63:127–33.

Jänig W, Kollmann W. The involvement of the sympathetic nervous system in pain. Artzneim-Forsch/Drug Res. 1984;34:1066–73.

Jänig W, Koltzenburg M. What is the interaction between the sympathetic terminal and the primary afferent fiber? In: Basbaum AI, Besson JM, editors. Towards a new pharmacotherapy of pain, Dahlem workshop reports. Chichester: Wiley; 1991. p. 331–52.

Jänig W, McLachlan EM. Organization of lumbar spinal outflow to the distal colon and pelvic organs. Physiol Rev. 1987;67:1332–404.

Jänig W, Schmidt RF. Report on the symposium: pathophysiological mechanisms of reflex sympathetic dystrophy, October 1990, Mainz, Germany. Clin Auton Res. 1991;1:73–4.

Jensen MP, Karoly P. Motivation and expectancy factors in symptom perception: a laboratory study of the placebo effect. Psychosom Med. 1991;53:144–52.

Jobling P, McLachlan EM, Jänig W, Anderson CR. Electrophysiological responses in the rat tail artery during reinnervation following lesions of the sympathetic supply. J Physiol. 1992;454:107–28.

Jung Y-H, Kim HK, Jeon SY, Kwon JM, Lee MD, Kim YC, Jang JH, Choi S-H, Lee J-Y, Kang D-H. Brain metabolites and peripheral biomarkers associated with neuroinflammation in complex regional pain syndrome using [11C]-(R)-PK11195 positron emission tomography and magnetic resonance spectroscopy: a pilot study. Pain Med. 2019;20:504–14.

Kalita J, Misra U, Kumar A, et al. Comparison of prednisolone in post-stroke complex regional pain syndrome. Pain Physician. 2016;19:565–74.

Kanoff RB. Intraspinal delivery of opiates by an implantable, programmable pump in patients with chronic, intractable pain of no-malignant origin. J Am Osteopath Assoc. 1994;94:487–93.

Kaufmann I, Eisner C, Richter P, Huge V, Beyer A, Chouker A, et al. Lymphocyte subsets and the role ofTH1/TH2 balance in stressed chronic pain patients. Neuro Immuno Mod. 2007;14:272–80.

Kawano M, Matsuoka M, Kurokawa T, et al. Autogenic training as an effective treatment for reflex neurovascular dystrophy: a case report. Acta Paediatr Jpn. 1989;31:500–3.

Kemler MA, van de Vusse AC, van den Berg-Loonen EM, Barendse GA, van Kleef M, Weber WE. HLA-DQ1 associated with reflex sympathetic dystrophy. Neurology. 1999;12:1350–1.

Kemler MA, Berendse GAM, van Kleef M, et al. Spinal cord stimulation in patients with chronic reflex sympathetic dystrophy. N Engl J Med. 2000;343:618–24.

Kesler RW, Saulsbury FT, Miller LT, Rowlingson JC. Reflex sympathetic dystrophy in children: treatment with transcutaneous electric nerve stimulation. Pediatrics. 1988;82:728–32.

Kessler A, Yoo M, Calsoff R. Complex regional pain syndrome: an updated comprehensive review. NeuroRehabilitation. 2020;47:253–64.

Kharkar DS, Ambady P, Yedatore V, et al. Intramuscular botulinum toxin A (BtxA) in complex regional pain syndrome. Pain Physician. 2011;14:311–6.

Kim SH, Chung JM. Sympathectomy alleviates allodynia in an experimental animal model for neuropathy inn the rat. Neurosci Lett. 1991;16:131–4.

Kim SH, Chung JM. An experimental model for peripheral neuropathy produced by segmental spinal nerve ligation in the rat. Pain. 1992;50:355–63.

Kim KD, De Salles AAF, Johnson JP, Ahn SS. Sympathectomy: open and thoracoscopic. In: Burchiel K, editor. Surgical management of pain. New York: Thieme; 2002. p. 688–700.

Kimura M, Suzuki N, Yoshino K. Reflex sympathetic dystrophy: a case report. Hattatsu. 1995;27:487–91.

Kimura T, Komatsu T, Hosoda R, Nishiwaki K, Shimada Y. Angiotensin-converting enzyme gene polymorphism in patients with neuropathic pain. In: Devor M, Rowbotham MC, Wiesenfeld-Hallin D, editors. Proceedings of the 9th World Congress on Pain. Progress in pain research and management, vol. 16. Seattle: IASP Press; 2000. p. 471–6.

Kingery WS. A critical review of controlled clinical trials for peripheral neuropathic pain and complex regional pain syndromes. Pain. 1997;73:123–39.

Kirklein JW, Chenoweth AI, Murphy F. Causalgia: a review of its characteristics, diagnosis and treatment. Surgery. 1947;21:321–42.

Klouchev AF, Frison A, Bauer T, Hardy P. Efficacy of vitamin C in preventing complex regional pain syndrome after wrist fracture: a systematic review and meta-analysis. Orthop Traumatol Surg Res. 2017;10:465–70.

Knudsen LF, Terkelsen AJ, Drummond PD, Birklein F. Complex regional pain syndrome: a focus on the autonomic nervous system. Clin Auton Res. 2019;29:457–67.

Kohr D, Singh P, Tschernatsch M, Kaps M, Pouokam E, Diener M, Kummer W, Birklein F, Vincent A, Goebel A, Wallukat G, Blaes F. Autoimmunity against the beta 2-Adrenergic receptor and muscarinic-2 receptor in complex regional pain syndrome. Pain. 2011;152:2690–700.

Koltzenburg M, Torbjörk HE. Pain and hyperalgesia in acute inflammatory and chronic neuropathic conditions. Lancet. 1995;345:1111.

Koltzenburg M, Häbler HJ, Jänig W. Functional reinnervation of the vasculature of the adult cat paw pad by axons originally innervating vessels in hairy skin. Neuroscience. 1995;767:245–52.

König S, Bayer M, Dimova V, Herrnberger M, Escolano-Lozano F, Bednarik J, Vickova E, Rittner H, Schlereth T, Birklein E. The serum protease network—one key to understand complex regional pain syndrome pathophysiology. Pain. 2019;160:1402–9.

König S, Engl C, Bayer M, Escolano-Lozano F, Rittner H. Substance P serum degradation in complex regional pain syndrome—another piece of the puzzle? J Pain. 2022;23:501–7.

Köster U, Stanton-Hicks M, Maihöfner C. What Paul Sudeck already suspected: from Sudeck's disease to complex regional pai syndrome. Schmerz. 2012;26:438–40.

Kozin F. Reflex sympathetic dystrophy syndrome. Bull Rheum Dis. 1986; 36: 1–8.

Kozin F. The reflex sympathetic dystrophy syndrome. Curr Opin Rheumatol. 1994;6:210–6.

Kozin F, Haughton V, Ruan L. The reflex sympathetic dystrophy syndrome in a child. J Pediatr. 1977;90:417–9.

Kozin F, Soin JS, Ryan LM, Carrera GF, Wortman RL. Bone scintigraphy in the reflex sympathetic dystrophy syndrome. Radiology. 1981;138:437–43.

Krämer HH, Hofbauer LC, Szalay G, Breimhorst M, Eberle T, Zieschang K, Rauner M, Schlereth T, Schreckenberger M, Birklein F. Osteoprotegerin: a new biomarker for impaired bone metabolism in complex regional pain syndrome? Pain. 2014;155:889–95.

Krämer HH, Eberle T, Üçyler N, Klonschinsky T, Müller LP, Sommer C, Birklein F. TNF-α in CRTPS and 'normal' trauma –significant differences between tissue and serum. Pain. 2011;152:285–90.

Kudlacek S, Schneider B, Woloszcuk W, Pietschmann P, Willvonseder R. Serum levels of osteoprotegerin increase with age in a healthy population. Bone. 2003;32:681–6.

Langston RG, Cowan RJ. Dupuytren's contracture. J Int Coll Surg. 1955;23:710.

Lee BH, Scharff L, Sethna NF, et al. Physical therapy and cognitive-behavioral treatment for complex regional pain syndromes. J Pediatr. 2002;141:135–40.

Lee Y, Lee CJ, Choi E, et al. Lumbar sympathetic block with botulinum toxin type A and type B for the complex regional pain syndrome. Toxins (Basel). 2018;10:164.

Lee JW, Nam H, Kim LE, Min JY, Ha S, Lee BY, Kim SY, Lee SJ, Kim EK, Yu SW. TLR4 (toll-like receptor 4) activation suppresses autophagy through inhibition of FOXO3 and impairs phagocytic capacity of microglia. Autophagy. 2019;15:753–70.

Lee HJ, Lee KH, Moon JY, Kim YC. Prevalence of autonomic nervous system dysfunction in complex regional pain syndrome. Reg Anesth Pain Med. 2021;46:198–202.

Lee B, Di Pietro F, Henderson LAS, Austin PJ. Altered basal ganglia infraslow oscillation and resting functional connectivity in complex regional pain syndrome. Neurosci Res. 2022;100(7):1487–505.

Lenz M, Uceyler N, Fretlöh J, et al. Local cytokine changes in complex regional pain syndrome type(CRPS 1) resolve after 6 months. Pain. 2013;154:2142–9.

Leriche R. De la causalgia comme envisege'e une ne'vrite du sympathique et son traitement par la de'nudation et l'excision des plexus nerveux periarteriels (in French). Presse Med. 1916;24:170–80.

Leriche R. La Chirurgie de la Doleur. Paris: Masson et Cie; 1949.

Lewis JS, Kersten P, McCabe CS, McPherson KM, Blake DR. Body perception disturbance: a contribution to pain in complex regional pain syndrome (CRPS). Pain. 2007;133:111–9.

Li X, Kenter K, Newman A, O'Brien S. Allergy/hypersensitivity reactions as a predisposing factor to complex regional pain syndrome 1 in orthopedic patients. Orthopedics. 2014;37:e286–91.

Li WW, Guo T-Z, Shi X, Birklein F, Schlereth T, Kingery WS, Clark JD. Neuropeptide regulation of adaptive immunity in the tibia fractur model of complex regional pain syndrome. J Neuroinflammation. 2018;15:105.

Livingstone WK. Pain mechanisms: a physiologic interpretation of causalgia and its related states. New York: Plenum Press (Reprint of the original ed published by McMillan, New York); 1976.

Loeb GE, White MW, Merzenich MM. Spatial cross-correlation. A proposed mechanism for acoustic pitch perception. Biol Cybern. 1983;47:149–63.

Logan DE, Carpino EA, Chiang G, et al. A day-hospital approach to treatment of complex regional pain syndrome: initial functional outcomes. Clin J Pain. 2012;28:766–74.

Loh L, Nathan PW, Schott GD. Pain due to lesions of central nervous system removed by sympathetic block. Br Med J. 1981;283:1026–8.

Low PA, Caskey PE, Tuck RR, Fealey RD, Dyck PJ. Quantitative sudomotor axon reflex test in normal and neuropathic subjects. Ann Neurol. 1983;14:573–80.

Lynch ME. Psychological aspects of reflex sympathetic dystrophy: a review of the adult and paediatic literature. Pain. 1992;49:337–347.

Maes M, Maes L, Schotte C, Cosyns PA. A clinical and biological validation of the DSM-III melancholia diagnosis in men: results of pattern recognition methods. J Psychiatr Res. 1992;26:183–96.

Maier C, Baron R, Tolle TR, Binder A, Birbaumer N, Birklein F, Gierthmuhlen J, Flor H, Geber C, Huge V, Krumova EK, Landwehrmeyer GB, Magerl W, Maihöfner C, Richter H, Rolke R, Scherens A, Schwartz A, Sommer C, Tronnier V, Uceyler N, Valet M, Wasner G, Treede RD. Quantitative sensory testing in the German Research Network on Neuropathic Pain (DFNS): somatosensory abnormalities in 1236 patients with different neuropathic pain syndromes. Pain. 2010;150:439–50.

Maihöfner C, Handwerker HO, Neundörfer B, Birklein F. Patterns of cortical reorganization in complex regional pain syndrome. Neurology. 2003;61:1707–15.

Maihöfner C, Handwerker HO, Neundörfer B, Birklein F. Cortical reorganization during recovery from complex regional pain syndrome. Neurology. 2004;63:693–701.

Maihöfner C, Handwerker HO, Neundorfer B, Birklein F. Mechanical hyperalgesia in complex regional pain syndrome: a role foer TNF-α? Neurology. 2005;65:311–3.

Maihöfner C, Baron R, DeCol R, Binder A, Birklein F, Deuschl G, Handwerker H, Schattschneider J. Motor system shows adaptive changes in complex regional pain syndrome. Pain. 2007;130(Pt 10):2671–87.

Mainka T, Bischoff FS, Baron R, Krumova EK, Nicolas V, Pennekamp W, Treede RD, Vollert J, Westermann A, Maier C. Comparison of muscle and joint pressure-pain thresholds in patients with complex regional pain syndrome and upper limb pain of other origin. Pain. 2014;155:591–7.

Mailis A Wade J. Profile of Caucasian women with possible genetic predisposition to reflex sympathetic dystrophy: a pilot study. Clin J Pain. 1994;10:210–7

Marinus J, Moseley GL, Birklein F, et al. Clinical features and pathophysiology of complex regional pain syndrome. Lancet Neurol. 2011;10:637–48.

Merzenich MM, Kaas JH, Wall J, Nelson RT, Sur M, Fellman D. Topographicb reorganization of somatosensory cortical areas 3b and1 in adult monkeys following restricted deafferentation. Neuroscience. 1983;8:33–35

Mascher WL. Über die Rolle des zentralen Nervensystems für die Entstehung des Sudeck-Syndroms. Nervenarzt. 1950;21:67–74.

Matak L, Tekus V, Bölcskei K, et al. Involvement of substance P in the antinociceptive effect of botulinum toxin type A: evidence from knockout mice. Neuroscience. 2017;358:137–45.

McCusker RH, McLaughlin PJ, Zagon LS. Immune-neural connections: how the immune system's response to infectious agents influences behavior. J Exp Biol. 2013;216:84–98.

McDonald MK, Ramanathan S, Tuati A, Zhou Y, Thanawala RU, Alexander GM, Sacan A, Ajit SK. Regulation of proinflammatory genes by the circulating microRNA hsa-miR-939. Sci Rep. 2016;6:30976.

Mekhail N, Deer TR, Poree L, Staats PS, Burton AW, Connolly AT, Karst E, Mehann YDS, Saweris Y, Levy RM. Cost-effectiveness of dorsal root ganglion stimulation or spinal cord stimulation for complex regional pain syndrome. Neuromodulation. 2021;24:708–18.

Merikangas KR, Frances A. Development of diagnostic criteria for headache syndromes: lessons from psychiatry. Cephalalgia. 1993;13(Suppl):34–8.

Merskey H, Bogduk N. Classification of chronic pain: descriptions of chronic pain syndromes and definitions of pain terms. 2nd ed. Seattle: IASP Press; 1994.

Mesaroli G, Hundert A, Birnie KSA, Campbell F, Stinson J. Screening and diagnostic tools for complex regional pain syndrome: a systemic review. Pain. 2021;152:1295–304.

Milligan ED Watkins LR. Pathological and protective roles of glia in chronic pain. Nat Rev Neurosci. 2009;10:23–36.

Misidou C, Papagoras C. Complex regional pain syndrome: an update. Mediterr J Rheumatol. 2019;30:16–25.

Mitchell SW, Morehouse GR, Keen WW. Gunshot wounds and other injuries of nerves. Philadelphia: JB Lippincott; 1864.

Morellini N, Finch PM, Goebe A, Drummond PD. Dermal nerve fibre and mast cell density and proximity of mast cells to nerve fibres in the skin of patients with complex regional pain syndrome. Pain. 2018;159:2021–9.

Mos D, Sturkenboom MC, Hugyen FJ. Current understandings on complex regional pain syndrome. Pain Pract. 2009;9:86–99.

Moseley GL. Graded motor imagery is effective for long standing complex regional pain syndrome: a randomized controlled trial. Pain. 2004;108:192–8.

Moseley GL. Graded motor imagery for pathologic pain: a randomized controlled trial. Neurology. 2006;67:2129–34.

Nahm FS, Park ZY, Nahm YC, Kim YC, Lee PB. Proteomic identification of altered cerebral proteins in the complex regional pain syndrome animal model. Biomed Res Int. 2014;2014:498410.

Nathan PW. On the pathogenesis of causalgia in peripheral nerve injuries. Brain. 1947;70:145–70.

Nathan PW, Spears TA. Some factors concerned in differential nerve block by local anesthetics. J Physiol. 1961;157:565–80.

Neumeister MW, Romanelli MR. Complex regional pain syndrome. Clin Plast Surg. 2020;47:305–10.

Nishiyama N, Cho SI, Kitagawa I, Saito H. Malonyl ginsenoside Rb1 potentiates nerve growth factor (NGF)-induced neurite outgrowth of cultured chick embryonic dorsal root ganglia. Biol Pharm Bull. 1994;17:509–13.

O'Connell NE, Wand BM, Mcauley J, Marston L, Moseley GL. Interventions for treating pain and disability in adults with complex regional pain syndrome—an overview of systematic reviews. Cochrane Database Syst Rev. 2013;6:CD009416. https://doi.org/10.1002/14651858.CD009416.pub2.

O'Connell NE, Wand BM, Gibson W, et al. Local anaesthetic sympathetic blockade for complex regional pain syndrome. Cochrane Database Syst Rev. 2016;7:CD004598.

Oaklander AL, Fields HL. Is reflex sympathetic dystrophy/complex regional pain syndrome type-I a small fiber neuropathy? Ann Neurol. 2009;65:629–38.

Oaklander AI, Rissmiller JG, Gelman LB, et al. Evidence of focal small-fiber axonal degeneration in complex regional pain syndrome-I (reflex sympathetic dystrophy). Pain. 2006;120:235–43.

Oakley JC, Weiner RL. Spinal cord stimulation for complex regional pain syndrome: a prospective study of 19 patients at two centers. Neuromodulation. 1999;2:47–50.

Ochoa JL. Reflex sympathetic dystrophy: a disease of medical understanding. Clin J Pain. 1992;8:363–36.

Ochoa JL. Pain mechanisms in neuropathy. Curr Opin Neurol. 1994;7:407–14.

Ochoa JS, Verdugo RJ. Neuropathic pain displayed by malingerers. J Neuropsychiatry Clin Neurosci. 2010;22:278–86.

Oerlemans HM, Goris RJA, de Boo T, Oostendorp RA. Do physical therapy and occupational therapy reduce the impairment percentage in reflex sympathetic dystrophy? Am J Phys Med Rehab. 1999a;78:533–9.

Oerlemans HM, Oostendorp RA, de Boo T, Perez RS, Goris RJ. Signs and symptoms in complex regional pain syndrome type I/ reflex sympathetic dystrophy: judgment of the physician versus objective measurement. Clin J Pain 1999b;15:224–3.

Oerlemans H, Oostendorp R, de Boo T, Laan L, Severens J, Goris R. Adjuvant physical therapy versus occupational therapy in patients with Reflex Sympathetic Dystrophy/Complex Regional Pain Syndrome Type One. Arch Phys Med Rehabil. 2000;81:49–56.

Ohmichi Y, Ohmichi M, Tashima R, Osuka K, Fuykushige K, Kanikowska D, Fukasawa Y, Yawo H, Tsuda M, Naito M, Nakano T. Physical disuse contributes to widespread chronic mechanical hyperalgesia, tactile allodynia and cold allodynia through neurogenic inflammation and spino-parabrachio-amygdaloid pathway activation. Pain. 2020;161:1808–23.

Olssen GL, Arner S, Hirsch G. Reflex sympathetic dystrophy in children. In: Tyler DC, Krane EJ, editors. Pediatric pain, vol. 15. New York: Raven Press; 1990. p. 323–31.

Orlova IA, Alexander GM, Rehman AQ, Sacan A, Graziano BJE, Schwartzman RJ, Ajit SE. MicroRNA modulation in complex regional pain syndrome. J Transl Med. 2011;10:195. https://doi.org/10.1186/1479-5876-9-195.

Osborne S, Farrell J, Dearman RJ, MacIver K, Naisbitt DJ, Moots RJ, et al. Cutaneous immuno-pathology of long-standing complex regional pain syndrome. Eur J Pain. 2015;19:1516–26.

Ott S, Maihöfner C. Signs and symptoms in 1043 patients with complex regional pain syndrome. J Pain. 2018;19:599–611.

Packham T, Holly J. Mechanism-specific rehabilitation management of complex regional pain syndrome: proposed recommendations from evidence synthesis. J Hand Ther. 2018;312:238–49.

Paget J. Clinical lecture on some cases of local paralysis. Med Times. 1864;1:331–2.

Pare A. Of the cure of wounds of the nervous system. In: The collected works of Ambroise Pare. New York: Milford House; 1634.

Parkitny L, McAuley JH, DiPietro F, Stanton TR, O'Connell NE, Marinus J, van Hilten JJ, Moseley GL. Inflammation in complex regional pain syndrome: a systemic review and meta-analysis. Neurology. 2013;80:106–17.

Paulignan Y, MacKenzie C, Marteniuk R, Jeannerod MN. Selective perturbation of visual input during prehension movements. 1. The effects of changing object position. Exp Brain Res. 1991;83:502–12.

Pearce JMS. Chronic regional pain and chronic pain syndromes. Spinal Cord. 2005;41:263–8.

Perez RS, Zuurmond WW, Bezemer PD, et al. The treatment of complex regional pain syndrome type I with free radical scavengers: a randomized controlled study. Pain. 2003;102:297–307.

Perez RS, Zollinger PE, Dijkstra PU, Thomassen-Hilgersom IL, Zuurmond WW, Rosenbrand KC, Geertzen JH. CRPS 1 task force. Evidence based guidelines for complex regional pain syndrome type 1. BMC Neurol. 2010;10:20.

Perez RSGM, Geertzen JHB, Dijkstra PU, Dirckx M, van Eijs F, Frölke JP. Updated guidelines: Complex regional pain syndrome type 1. Netherland s Society of Anaesthesiologists/ Netherlands Society of Rehabilitation Specialists. 2014. http://pdver.atcomputing.nl/pdf/ Executive_summary_updated_guidelines_CRPS-1-2014.pdf.

Perl E. Causalgia, pathological pain, ands adrenergic receptors. Proc Natl Acad Sci U S A. 1999;96:7664–7.

Pette H. Das Problem der wechselseitigen Beziehungen zwischen Sympathicus und Sensibilität. Dtsch Zschr Nervenheilkunde. 1927;100:143–1148.

Pilowsky I. Abnormal illness behavior. Br J Med Psychol. 1969;42:347–51.

Pilowsky I, Spence N, Cobb J, Katsikitis M. The illness behavior Questionaire as an aid to clinical assessment. Gen Hosp Psychiatry. 1984;6:123–30.

Pleger B, Tegenhoff M, Ragert P, Föester A-F, Dinse HR, Schwenkreis P, Nicolas V, Maier C. Sensorimotor retuning (corrected) in complex regional pain syndrome parallels pain reduction. Ann Neurol. 2005;57:426–9.

Poplawski ZJ, Wiley AM, Murray JF. Post-traumatic dystrophy of the extremities. J Bone Joint Surg Am. 1983;65:642–55.

Price DD, Bennett GJ, Rafii A. Psychophysical observations on patients with neuropathic pain relieved by a sympathetic block. Pain. 1989;36:273–88.

Price DD, Long S, Huitt C. Sensory testing of pathophysiological mechanisms of pain in patients with reflex sympathetic dystrophy. Pain. 1992;49:163–73.

Rahn KA, McLaughlin PJ, Zagon LS. Prevention and diminished expression of experimental autoimmune encephalomyelitis by low dose naltrexone (LDN) or opioid growth factor (OGF) for an extended period: therapeutic implications for multiple sclerosis. Brain Res. 2011;1381:243–53.

Ramachandran VS, Herstein W. The perception of phantom limb. The D. O. Hebb lecture. Brain. 1998;121:1603–30.

Ramanathan S, Douglas SR, Alexander GM, Shenoda BB, Barratt JE, Aradillas E, Sacan A, Ajitn SK. Exosome microRNA signatures in patients with complex regional pain syndrome undergoing plasma exchange. J Transl Med. 2019;17:81.

Rauck RL, Eisenach JC, Jackson K, Young LD, Southern J. Epidural clonidine treatment of refractory reflex sympathetic dystrophy. Anesthesiology. 1993;79:1163–9.

Reidl B, Beckmann T, Neundorfer B, Handwerker HO, Birklein F. Autonomic failure after stroke—is it indicative for pathophysiology of complex regional pain syndrome? Acta Neurol Scand. 2001;103:27–34.

Reinhold AK, Kindl G-K, Dietz C, Scheu N, Mehling K, Brack A, Birklein F, Rittner HL. Molecular and clinical markers of pain relief in complex regional pain syndrome: an observational study. Eur J Pain. 2022;27:278–88.

Revicki DA, Celia DF. Health status assessment from the twenty-first century: item response theory, item banking and computer adaptive testing. Qual Life Res. 1997;5:595–600.

Ribbers GM, Osterhuis WP, van Limbeek J, de Metz M. Reflex sympathetic dystrophy: is the immune system? Arch Phys Med Rehabil. 1998;79:1549–52.

Richards RL. Causalgia: a centennial review. Arch Neurol. 1967;16:339–50.

Richlin DM, Carron H, Rowlingson JC, Sussman MD, Baugher WH, Goldner RD. Reflex sympathetic dystrophy: successful treatment by transcutaneous nerve stimulation. J. Pediatr. 1978;93:84–6.

Ritz BW, Alexander GM, Noguss S, Perrault MJ, Peterlin BL, Grothusen JR, et al. Elevated blood levels of inflammatory monocytes (CD14 + CD 16+) in patients with complex regional pain syndrome. Clin Exp Immunol. 2011;164:108–17.

Roberts WJ. A hypothesis on the physiological basis for causalgia and related pains. Pain. 1986;24:297–311.

Roberts WJ, Elardo SM. Sympathetic activation of unmyelinated mechanoreceptors in cat skin. Brain Res. 1985;339:123–5.

Roberts WJ, Foglesong ME. Spinal recordings suggest that wide-dynamic-range neurons mediate sympathetically maintained pain. Pain. 1988;34(3):289–304. https://doi.org/10.1016/0304-3959(88)90125-X. PMID: 3186277.

Robinson JN, Sandom J, Chapman PT. Efficacy of pamidronate in complex regional pain syndrome type I. Pain Med. 2004;5:276–80.

Rommel O, Gehling M, Dertwinkel R, Witscher K, Zenz M, Malin JP, Jänig W. Hemisensory impairment in patients with complex regional pain syndrome. Pain. 1999;80(1–2):95–101.

Rommel O, Malin JP, Zenz M, Jänig W. Quantitative sensory testing, neurophysiological and psychological examination in patients with complex regional pain syndrome and hemisensory deficits. Pain. 2001;93:279–93.

Rubin LA, Snow KM, Kurman CC, Nelson DL, Keystone EC. Serial levels of soluble interleukin 2 recfeptor in the peripheral blood of patients with rheumatoid arthritis: correlation with disease activity. J Rheumatol. 1990;17:597–602.

Russo MA, Santarelli DM. AS novel compound analgesic cream (ketamine, pentoxifylline, clonidine, DMSO) for complex regional pain syndrome patients. Pain Pract. 2016;16:E14–20.

Russo MA, Fiore NT, van Vreden C, Bailey D, Santarelli DM, McGuire HM, Fazekas de St Groth B, Austin PJ. Expansion and activation of distinct central memory T-lymphocyte subsets in complex regional pain syndrome. J Neuroinflammation. 2019;16:63. https://doi.org/10.1186/s12974-019-144.

Russo MA, Georgius P, Pires AS, Heng B, Allwright M, Guennewig B, Santarelli DM, Bailey D, Fiore NT, Tan VX, Latini A, Guillemin GJ, Austin PJ. Novel immune biomarkers in complex regional pain syndrome. J Neuroimmunol. 2020;347:577330. https://doi.org/10.1016/j.jneuroimmun.2020.577330.

Sandroni P, Low PAS, Ferrer T, Opfer-Gehrking TL, Wilson PR. Complex regional pain syndrome (CRPS): prospective study and laboratory evaluation. Clin J Pain 1998;14:282–9.

Sandroni P, Benrud-Larson LM, McClelland RL, Low PA. Complex regional pain syndrome type I: incidence and prevalence in Olmstead County, a population-based study. Pain. 2003;106:199–207.

Sator-Katzenschlager S, Deusch E, Maier P, Spacek A, Kress HG. The long-term antinociceptive effect of intrathecal S(+)- ketamine in. a patient with established morphine tolerance. Anesth Analg. 2001;93(4):1032–4.

Schattschneider J, Wenzelburger R, Deuschl G, Baron R. Kinematic analysis of the upper extremity in CRPS. In: Harden RN, Baron R, Jänig W, editors. Complex regional pain syndrome. Progress in pain research and management, vol. 22. Seattle: IASP Press; 2001. p. 119–28.

Schiller JE. Reflex sympathetic dystrophy of the foot and ankle in children and adolescents. J Am Podiatr Med Assoc. 1989;79:545–51.

Schlereth T, Birklein F. Mast cells: source of inflammation in complex regional pain syndrome. Anesthesiology. 2012;118:756–7.

Schwartzman R. New treatment for reflex sympathetic dystrophy. N Engl J Med. 2000;343:654–6.

Schwartzman RJ, McLellan TL. Reflex sympathetic dystrophy: a review. Arch Neurol. 1987;44:555–61.

Schwartzman RJ, Liu JE, Smullens SN, Hyslop T, Tahmoush AJ. Long-term outcome following sympathectomy for complex regional pain syndrome type 1 (RSD). J Neurol Sci. 1997;150(2):149–52. https://doi.org/10.1016/s0022-510x(97)00078-6. PMID: 9268243.

Schwartzman RJ, Erwin KL, Alexander M. The natural history of complex regional pain syndrome. Clin J Pain. 2009;25:273–80.

Seltzer Z, Dubner R, Shir Y. A novel behavioral model of neuropathic pain disorders produced in rats by partial sciatic nerve injury. Pain. 1900;43:205–18.

Shenker N, Goebel A, Rockett M, Batchelor J, Jones GT, Parker R, McCabe C. Establishing the characteristics for patients with complex regional pain syndrome: the value of the CRPS-UK registry. Br J Pain. 2015;9:123–8.

Sherry DD, Weisman R. Psychological aspects of childhood reflex sympathetic dystrophy. Pediatrics. 1988;81:572–8.

Shi T, Apkarian P. Morphology of thalamocortical neurons projecting to the primary sensory cortex and their relationship to spinothalamic terminals in the squirrel monkey. J Comp Neurol. 1995;36:1–24.

Shu XQ, Mendell LM. Neurotropins and hyperalgesia. Proc Natl Acad Sci U S A. 1999;96:7693–6.

Siegel S. Opioid expectation modifies opioid effects. Fed Proc. 1982;41:2339–43.

Siegel SM, Oaklander AL. Needlestick injury in rat models symptoms of complex regional pain syndrome. Anesth Analg. 2007;105:1820–9.

Singh G, Willen S, Boswell M, Janata J, Chelimsky T. The value of interdisciplinary pain management in complex regional pain syndrome type one: a prospective outcome study. Pain Physician. 2004;7:203–9.

Soin A, Soin Y, Buenaventura R, Ferguson K, Atluri S, Sachdeva H, Sudarshan G, Akbik H, Italiano J. Low-dose naltrexone use for patients with chronic regional pain syndrome: a systematic literature review. Pain Physician. 2021;24:E393–406.

Sorge RE, LaCroix-Fralish ML, Tuttle AH, Sotocinal SG, Austin J-S, Ritchie J, Chanda ML, Graham AC, Topham L, Beggs S, Salter MW, Mogul JS. Spinal cord Toll-like receptor 4 mediates inflammatory and neuropathic hypersensitivity in male but not female mice. J Neurosci. 2011;31:15450–4.

Stanton RP, Malcolm JR, Wesdock KA, Singsen BH. Reflex sympathetic dystrophy in children: an orthopedic perspective. Orthopedics. 1993;16:773–9.

Stanton-Hicks M. Complex regional pain syndrome. Anesthesiol Clin North Am. 2003;21:733–44.

Stanton-Hicks M. Plasticity of complex regional pain syndrome (CRPS) in children. Pain Med. 2010;11:1216–3.

Stanton-Hicks M. CRPS: what's in a name? Taxonomy. Epidemiology, neurologic, immune and auto-immune considerations. Reg Anesth Pain Med. 2019;44:376–87.

Stanton-Hicks M, Jänig W, Hassenbusch S, et al. Reflex sympathetic dystrophy: changing concepts and taxonomy. Pain. 1995;63:127–33.

Stanton-Hicks M, Baron R, Boas R, Gordh T, Harden N, Hendler N, Koltzenburg M, Raj P, Wilder R. Complex regional pain syndromes: guidelines for therapy. Clin J Pain. 1998;14:155–66.

Stanton-Hicks M, Burton AW, Bruehl SP, Carr DB, Harden RN, Hassenbusch SJ, Lubenow TR, Oakley JCV, Racz GB, Raj P, Rauck RL, Rezai AR. An updated interdisciplinary clinical pathway for CRPS: report of an export panel. Pain Pract. 2002;2:1–16.

Strauss S, Barby S, Hartner J, Neumann N, Moseley GL. Modifications in fMRI representation of mental rotation following a 6 week graded motor imagery training in chronic CRPS patients. J Pain. 2020;22:680–91.

Stude P, Enax-Krumova EK, Zenz M, Lissek S, Nicolas V, Peters S, Westerman A, Tegenhoff M, Maier C. Local anesthetic sympathectomy restores FMRI cortical maps in CRPS I after upper extremity stellate ganglion blockade: a prospective case study. Pain Physician. 2014;17:E637–44.

Sudeck P. Über die akute entzündliche Knochenatrophie. Arch Klin Chir. 1900;62:147–56.

Sudeck P. Die sogenannte akute Knochenatrophie als Entzündingsvorgang. Chirurg. 1942;14:449–58.

Sunderland S, Kelly M. The painful sequelae of injuries to peripheral nerves. Aust New Zeal J Surg. 1948;18:75–118.

Tang C, Li J, Tai WL, Yao W, Zhao B, Hong J, Shi S, Wang S, Xia Z. Sex differences in complex regional pain syndrome type I (CRPS-I) in mice. J Pain Res. 2017;31:1811–9.

Teasell RW, Arnold JMO. Alpha-1 adrenoceptor hyperresponsiveness in three neuropathic pain states: complex regional pain syndrome 1, diabetic peripheral neuropathic pain and central pain states following spinal cord injury. Pain Res Manag. 2004;9:89–97.

Tekus V, Hajna ZB, Nioirbely E, Marcovics A, Bagoly T, Szolccsanyi J, Thompson V, Kemeny A, Helyes Z, Goebel A. A CRPS-IgG-transfer-trauma model reproducing inflammatory and positive signs associated with complex regional pain syndrome. Pain. 2014;155:299–308.

Ter Borg EJ, Horst G, Limburg PC, Kallenberg CG. Changes in plasma levels of interleukin-2 receptor in relation to disease exacerbations and levels of anti-dsDNA and complement in systemic lupus erythematosus. Clin Exp Immunol. 1990;82:21–6.

Terkelsen A, Giertmühlen G, Petersen LJ, Knudsen L, Christensen NJ, Kehr J, Yoshitake T, Madsen CS, Wasner G, Baron R, Jensen TS. Cutaneous noradrenaline measured by microanalysis in complex regional pain syndrome during whole-body cooling and heating. Exp Neurol. 2013;247:456–65.

Thimineur M, Sood P, Kravitz J, Kitaj M. Central nervous system abnormalities in complex regional pain syndrome (CRPS): clinical and quantitative evidence of medullary dysfunction. Clin J Pain. 1998;14:256–67.

Toljan K, Vrooman B. Low-dose naltrexone (LDN)—review of therapeutic utilization. Med Sci (Basil). 2018;6:82. https://doi.org/10.3390/medsci6040082.

Torebjörk E, Wahren L, Wallin G, Hallin R, Koltzenburg M. Noradrenaline-evoked pain in neuralgia. Pain. 1995;63:11–20.

Touzet P, D'Ornano P, Chaumien JP, Prieur SAM, Tigault P. Misleading forms of algoneurodystrophy in children and adolescents apropos of 19 cases. Annales de Pediatrie. 1991;38:673–381.

Tracey DJ, Cunningham JE, Romm MA. Peripheral hyperalgesia in experimental neuropathy: mediation by alpha 2-adrenoceptors on post-ganglionic sympathetic terminals. Pain. 1995;60:317–27.

Treede RD, Davis KD, Campbell JN, Raja SN. The plasticity of cutaneous hyperalgesia during sympathetic ganglion blockade in patients with neuropathic pain. Brain. 1992;115:607–21.

Treharne G, Lyons A, Hale E, Douglas K, Kitas G. Compliance' is futile but 'concordance' between rheumatology patients and health professionals attainable ? Rheumatology. 2006;45:1–5.

Tsuda M, Shigimoto-Mogami Y, Koizumi S, Mizokoshi A, Kohsaka S, Salter MW, Inoue K. P2X4 receptors induced in spinal microglia gate tactile allodynia after nerve injury. Nature. 2003;424:778–83.

Turk DC, Melzack R. The measurement of pain and the assessment of people experiencing pain. In: Turk DC, Melzack R, editors. Handbook of pain assessment. New York: Guilford Press; 2001. p. 1–11.

Tuttle JB, Etheridge R, Creedon DJ. Receptor-mediated stimulation and inhibition of nerve growth factor secretion by vascular smooth muscle. Exp Cell Res. 1993;208:350–61.

Ulmer JL, Mayfield FH. Causalgia; a study of 75 cases. Surg Gynecol Obstet. 1946;83:789–96.

Urits I, Shen AH, Joners MR, Viswanath O, Kaye AD. Complex regional pain syndrome, current concepts and treatment options. Curr Pain Headache Rep. 2018;22(2):10.

Van Buyten J-P, Smet I, Liem L, Russo M, Huygen F. Stimulation of dorsal root ganglia for the management of complex regional pain syndrome: a prospective case series. Pain Pract. 2015;15:208–16.

Van de Meent H, Oerlemans M, Bruggeman A, Klomp F, van Dongen R, Oostendorp R, Frölke JP. Safety of "pain exposure" physical therapy in parents with complex regional pain syndrome type 1. Pain. 2011;152:1431–8.

Van den Beek WJT, Schwartzman RJ, Van Nes SI, Delhaas EM, Van Hilten JJ. Diagnostic criteria used in studies of reflex sympathetic dystrophy. Neurology. 2002;58:522–6.

Van der Kloot WA, Oostendoorp RA, van der Meij J, van den Heuvel J. The Dutch version of the McGill pain questionnaire: a reliable pain questionnaire. Ned Tijdschr Geneeskd. 1995;139:669–73.

Van der Kloot WA, Oostendoorp RA, van der Meij J, van den Heuvel. The Dutch version of the McGill pain questionnaire: a reliable pain questionnaire. Ned Tijdschr Geneeskd. 1995b;139:669–73.

Van der Laan L, Goris RJA. Sudeck-syndrom: hates Sudeck recht? Unfallchirurg. 1997;100:90–9.

Van der Laan L, ter Laak HJ, Gabreels-Festen AG, Abreels F, Goris RJA. Complex regional pain syndrome type I (RSD): pathology of skeletal muscle and peripheral nerve. Neurology. 1998;51:20–5.

Van Hilten BJ, van de Beek WJT, Hoff JI, Voormolen JHC, Delhaas EM. Intrathecal baclofen for the treatment of dystonia in patients with reflex sympathetic dystrophy. N Engl J Med. 2000;343:625–30.

Van Houdenove B. Neuro-algodystrophy: a psychiatrist's view. Clin Rheumatol. 1986;5:399–406.

Varenna M, Adami S, Rossini M, et al. Treatment of complex regional pain syndrome type I with Neridronate: a randomized double-blind, placebo-controlled study. Rheumatology. 2013;52:534–42.

Veldman P, Reynen H, Arntz I, Goris R. Signs and symptoms of reflex sympathetic dystrophy: prospective study of 829 patients. Lancet. 1993;342:1012–6.

Verdugo RJ, Ochoa JL. "Sympathetically maintained pain." I. Phentolamine block questions the concept. Neurology. 1994;44:1003–10.

Verdugo RJ, Ochoa JL. Reversal of hypoesthesia by nerve block, or placebo: a psychologically mediated sign in chronic pseudoneuropathic pain. J Neurol Neurosurg Psychiatry. 1998;65:196–203.

Verdugo MS, Arias B, Ibanez A, Schalock RL. Adaptation and psychometric properties of the Spanish version of the Supports Intensity Scale (SIS).2010. Am J Intellect Dev Disabil. 2010;115:496–503.

Visnjevac O, Costandi S, Patel BA, Azer G, Agarwal P, Bolash R, Mekhail NA. A comprehensive outcome-specific review of the use of spinal cord stimulation for complex regional pain syndrome. Pain Pract. 2017;17:533–45.

Vlaeyen JW, Kole-Snijders AM, Boeren RG, van Eek H. Fear of movement/(re)injury in chronic low back pain and relation to behavioral performance. Pain. 1999;62:363–72.

Vorselaars AD, Van Moorsel CH, Zanen P, Ruven HJ, Claessen AM, Velzen-Blad H, et al. Respir Med. 2015;109:279–85.

Vranken JH, Van der Vegt MH, Kaql JE, Kruis MR. Treatment of neuropathic cancer pain with continuous intrathecal administration of S (+)-ketamine. Acta Anaesthesiol Scand. 2004;48:249–52.

Wakisaka S, Kajander KC, Bennett GJ. Abnormal skin temperature and abnormal sympathetic vasomotor innervation in an experimental painful peripheral neuropathy. Pain. 1991;46:293–313.

Wallin G, Torebjörk E, Halim R. Preliminary observations on the pathophysiology of hyperalgesia in the causalgic syndrome. In: Zotterman Y, editor. Sensory functions of the skin in primates, Wennergren International Symposium, vol. 27. Oxford: Pergamon Press; 1976. p. 489–502.

Wang X, Zhang Y, Peng Y, Hutchinson MR, Rice KC, Yin H, Watkins LR. Pharmacological characteristics of the opioid inactive isomers (+)-naltrexone and (+)-naltrexone as antagonists of toll-like receptor 4. Br J Pharmacol. 2016;173:856–69.

Wang J, Zheng X, Liu B, Yin C, Chen R, Li X, Li Y, Nie H, Zeng D, He X, Jiang Y, Fang Y, Liu B. Electroacupuncture alleviates mechanical allodynia of a rat model of CRPS-I and modulates gene expression profiles in dorsal root Anglia. Front Neurol. 2020;11:580997.

Warren JR, Marshall B. Unidentified curved bacilli on gastric epithelium in active chronic gastritis. Lancet. 1983;1:1273–5.

Wasner G, Heckmann C, Maier C, Baron R. Vascular abnormalities in acute reflex sympathetic dystrophy (CRPS I): complete inhibition of sympathetic nerve activity with recovery. Arch Neurol. 1999;56:613–20.

Wasner G, Drummond P, Birklein F, Baron R. The role of the sympathetic nervous system in autonomic disturbance and "sympathetically maintained pain" in CRPS. In: Harden RN, Baron R, Jänig W, editors. Complex regional pain syndrome, progress in pain research and management, vol. 22. Seattle: IASP Press; 2001. p. 89–108.

Watson HK, Carlson L. Treatment of reflex sympathetic dystrophy of the hand with an active "stress loading" program. J Hand Surg. 1987;12A:779–85.

Weber M, Birklein F, Neundorfer B, Schmelz M. Facilitated neurogenic inflammation in complex regional pain syndrome. Pain. 2001;91:251–7.

Wesseldijk F, Huygen FJ, Heijmans-Antonissen C, Niehof SP, Zijlstra FJ. Tumor necrosis factor-alpha and interleukin-6 are not correlated with the characteristics of Complex Regional Pain Syndrome type 1 in 66 patients. Eur J Pain. 2008;12:716–21.

Weiner RL. The future of peripheral nerve neurostimulation. Neurol Res. 2000;22:299–303.

Wertli M, Bachman L, Weiner S, Brunner F. Prognostic factors in complex regional pain syndrome 1.: a systematic review. J Rehabil Med. 2013;45:225–31.

White JC, Sweet WH. Pain and the neurosurgeon: a forty-year experience. Springfield: CC Thomas; 1969.

White FA, Bhangoo SK, Miller RJ. Chemokines: integrators of pain and inflammation. Nat Rev Drug Discov. 2005;4:834–44.

Wickman JR, Lug X, Li W, Jean-Toussaint R, Sahbaie P, Sacan A, Clark JD, Ajit SX. Circulating micro-RNA's from the mouse tibia fracture model reflect the signature from patients with complex regional pain syndrome. Pain Rep. 2021;6:e950. https://doi.org/10.1097/PR9.

Wilder RT, Berde CB, Wolohan M, Vieyra MA, Masek BJ, Micheli LJ. Reflex sympathetic dystrophy in children. Clinical characteristics and follow-up of seventy patients. J Bone Joint Surg. 1992;74:910–9.

Wilson PR, Low PA, Bedder MD, Covington EC, Rauck RL. Diagnostic algorithm for complex regional pain syndromes. In: Jänig W, Stanton-Hicks M, editors. Reflex sympathetic dystrophy: a reappraisal. Progress in pain research and management, vol. 6. Seattle: IASP Press; 1996. p. 93–105.

Wong GY, Wilson PR. Classification of complex regional pain syndromes. New concepts. Hand Clin. 1997;13:319–25.

Xanthos DN, Sandkühler J. Neurogenic inflammation: inflammatory CNS reactions in response to neuronal activity. Nat Rev Neurosci. 2014;15:43–53.

Xanthos DM, Bennet GJ, Coderre TJ. Norepinephrine-induced nociception and vasoconstrictor hypersensitivity in rats with chronic post-ischemia pain. Pain. 2008;137:640–51.

Yarnitsky D, Fowler CJ. Quantitative sensory testing. In: Osselton J, editor. Neurophysiology. London: Butterworth; 1995.

Yarnitsky D, Sprecher E. Thermal testing: normative data and repeatability for various test algorithms. J Neurol Sci. 1994;39:39–45.

Yu X, Wang Y, Zhou H, Yang W, Guan X. Toll-like receptor 7 promotes the apoptosis of THP-1-derived macrophages through the CHOP-dependent pathway. Int J Mol Med. 2014;34:886–93.

Ziegler-Heitbrock L, Ancuta P, Crowe S, Dalod M, Grau V, Hart DN. Nomenclature of monocytes and dendritic cells in blood. Blood. 2010;116:74–80.

Zollinger PE, Tuinebreijer WE, Kreis RW, Breederveld RS. Effects of vitamin C on frequency of reflex sympathetic dystrophy in wrist fractures: a randomized trial. Lancet. 1996;354:2025–8.

Zuurmond WWA, Landijk PNJ, Bezemer PD, et al. Treatment of acute reflex sympathetic dystrophy with DMSO 50% in a fatty cream. Acta Anaesthesiol Scand. 1996;40:363–36.

Index